大地测量学基础(双语教程)

Foundation of Geodesy: Bilingual Reading Book

张西光　吕志平　李　健　曲云英　李艳霞　**编著**

测绘出版社

·北京·

ⓒ 张西光　吕志平　李　健　曲云英　李艳霞　2011
所有权利(含信息网络传播权)保留，未经许可，不得以任何方式使用。

内 容 提 要

本书为测绘类专业本科《大地测量学基础》(普通高等教育"十一五"国家级规划教材，吕志平，乔书波编著，测绘出版社，2010.3)的英语辅助教材，目的是使学生在学习《大地测量学基础》课程的同时，学习掌握该课程的基本概念、基本理论在英文中的表达和描述，为以后阅读英文专业文献，撰写专业论文打下良好的基础。

本书可作为高等院校测绘类各专业本科生的通用教材，对于从事与测绘工程有关的技术人员也是一本值得推荐的基础性参考书。

图书在版编目(CIP)数据

大地测量学基础：双语教程/张西光等编著.－－北京：测绘出版社，2011.10 (2024.1重印)
ISBN 978-7-5030-2433-7

Ⅰ．①大… Ⅱ．①张… Ⅲ．①大地测量学－双语教学－高等学校－教材 Ⅳ．①P22

中国版本图书馆 CIP 数据核字(2011)第 196203 号

责任编辑　田　力	封面设计　李　伟	责任校对　董玉珍　李　艳	责任印制　陈姝颖

出版发行　测绘出版社	电　话　010－68530735(发行部)
地　　址　北京市西城区三里河路50号	010－68531363(编辑部)
邮政编码　100045	网　址　https://chs.sinomaps.com
电子邮箱　smp@sinomaps.com	经　销　新华书店
成品规格　184mm×260mm	印　刷　北京建宏印刷有限公司
印　　张　11	字　数　316千字
版　　次　2011年10月第1版	印　次　2024年1月第3次印刷
印　　数　2301—2800	定　价　36.00元
书　　号　ISBN 978-7-5030-2433-7	

本书如有印装质量问题，请与我社发行部联系调换。

前　言

通过阅读专业英文文献，获取知识，进而进行科学研究和学术交流，是英语教育的一项重要目标。从通用英语的学习，过渡到使用英语学习专业课程，并非简单、自然完成的，通常需要一个过程。大地测量学基础课程通常开设在第四或第五学期，编写一本双语教程，加强学生在这方面的训练，顺利实现这一过渡，是一件很有意义的工作。

编写本书的目的在于，使学生在学习《大地测量学基础》（普通高等教育"十一五"国家级规划教材，吕志平，乔书波编著，测绘出版社，2010.3）课程的同时，学习掌握该课程的基本概念、基本理论在英文中的表达和描述，为以后阅读英文专业文献，撰写英文专业论文打下良好的基础。这也是本课程与专业英语课程的教学内容定位不同之处。

本书的英文教学内容，取材于英、美和加拿大等国的测绘类专著、科技论文和科技词典，以选取原文为主，部分内容略加变动，符号改为我国惯用符号。在选材时，除了注重内容的原汁原味，还注重把握下面四条原则。第一，选取的内容能够激发学生阅读专业英文文献的兴趣，例如，给出了大地测量学的四个定义，从不同角度出发阐释，引发学生思考；关于大地测量学历史的回顾，与中文教材上的内容略有不同，从不同侧面审视大地测量学的发展。第二，选取的内容与大地测量学基础课程的内容既互为对照，又互为补充，可作为课外学习的延伸，例如，大地问题解算的逐点积累法，本书选择了编程实现该方法的内容；再如，关于 WGS-84 坐标系和 ITRF 坐标系的内容，本书选取了较多材料，便于课后扩展学习。第三，所选专业内容在科技文献中表达的典型性，从中学习各种表达方式和技巧，例如，选取的内容包括术语的定义，公式推导，某领域综述及总结，等等。第四，选取的内容很多出自著名大地测量学家之手，其科学性自不必多言，更重要的是学习到他们对问题的认识，例如，关于时间系统的描述极为简洁；关于板块构造运动的描述特别形象，给人以深刻印象。此外，为方便学生自学，增加了专业词汇表和参考译文。

本书在编写的过程中，得到信息工程大学测绘学院训练部张晓森部长、郭延斌主任，一系李广云主任和一系大地测量教研室全体同仁的大力支持。宋力杰教授、柴洪洲教授审阅了全书，并提出了许多宝贵意见，刘长建副教授、赵冬青副教授和乔书波副教授在搜集素材，文献翻译中提供了很大的帮助，在此表示衷心感谢。本书的出版还要感谢测绘学院"2110 学科建设"的资助。

由于编者水平有限，经验不足，书中存在的缺点和错误，恳请读者批评、指正，以促进本课程建设的不断发展。

Contents

1. Introduction 1
   - 1.1 Definition of Geodesy 1
   - 1.2 Classification of Geodesy 1
   - 1.3 Aim of Geodesy 2
   - 1.4 Historical Introduction 2
   - 1.5 Recent Developments of Space Geodesy 3
   - Vocabulary 5

2. Reference System 6
   - 2.1 Time Systems 6
   - 2.2 International Earth Rotation Service 7
   - 2.3 Celestial Reference System 8
   - 2.4 Terrestrial Reference System 11
   - 2.5 Gravity Field Related Reference Systems 14
   - Vocabulary 16

3. Methods of Measurement in Geodesy 18
   - 3.1 Terrestrial Geodetic Measurements 18
   - 3.2 Satellite Observations 20
   - 3.3 Gravity Measurements 24
   - 3.4 Astronomic Measurements 25
   - Vocabulary 26

4. Geodetic Networks 27
   - 4.1 Horizontal Control Networks 27
   - 4.2 Vertical Control Networks 29
   - 4.3 Three-dimensional Networks 30
   - 4.4 Gravity Networks 30
   - Vocabulary 31

5. The gravity field of the Earth 32
   - 5.1 Fundamentals of the Gravity Field 32
   - 5.2 Level Surfaces and Plumb Lines 33

  5.3 Spherical Harmonic Expansion of the Gravitational Potential ………………… 34
  5.4 The Geoid ……………………………………………………………………… 34
  5.5 Height Systems ………………………………………………………………… 36
  Vocabulary ………………………………………………………………………… 38
6 Reference Ellipsoid and Geodetic Coordinate System ………………………… 39
  6.1 The Rotational Ellipsoid ……………………………………………………… 39
  6.2 Curvature ……………………………………………………………………… 44
  6.3 Geodesics ……………………………………………………………………… 47
  6.4 Reductions to the Ellipsoid …………………………………………………… 49
  6.5 Direct and Inverse Geodetic Problems ……………………………………… 60
  Vocabulary ………………………………………………………………………… 64
7 Map Projection ………………………………………………………………………… 65
  7.1 Introduction of Map Projection ……………………………………………… 65
  7.2 Projection Mathematics ……………………………………………………… 67
  7.3 Conformal Map Projection …………………………………………………… 68
  7.4 Transverse Mercator Projection ……………………………………………… 71
  Vocabulary ………………………………………………………………………… 76
8 Establishment of Geodetic Coordinate System ………………………………… 77
  8.1 Datums ………………………………………………………………………… 77
  8.2 Geodetic Datum ……………………………………………………………… 78
  8.3 Coordinate Transformation and Datum Shifts ……………………………… 81
  8.4 Orientation of Astrogeodetic Systems, Best Fitting Ellipsoids …………… 86
  8.5 Three-dimensional Transformations ………………………………………… 87
  8.6 World Geodetic System 1984 ………………………………………………… 90
  8.7 ITRS and Its Realization ……………………………………………………… 91
  Vocabulary ………………………………………………………………………… 98
References ………………………………………………………………………………… 99
Vocabulary ………………………………………………………………………………… 100

参考译文

1 概　论 ……………………………………………………………………………… 104
  1.1 大地测量学的定义 …………………………………………………………… 104
  1.2 大地测量学的分类 …………………………………………………………… 104

1.3	大地测量学的任务	104
1.4	历史简介	105
1.5	空间大地测量学的近来发展	105

2 参考系统 ·········· 106
- 2.1 时间系统 ·········· 106
- 2.2 国际地球自转服务（IERS） ·········· 108
- 2.3 天球参考系 ·········· 108
- 2.4 地球参考系 ·········· 110
- 2.5 与重力场有关的参考系 ·········· 112

3 大地测量中的观测技术 ·········· 113
- 3.1 地面大地测量观测 ·········· 113
- 3.2 卫星测量 ·········· 115
- 3.3 重力测量 ·········· 117
- 3.4 天文测量 ·········· 118

4 大地控制网 ·········· 119
- 4.1 平面控制网 ·········· 119
- 4.2 高程控制网 ·········· 120
- 4.3 三维控制网 ·········· 121
- 4.4 重力控制网 ·········· 121

5 地球重力场 ·········· 122
- 5.1 重力场基础 ·········· 122
- 5.2 水准面和铅垂线 ·········· 123
- 5.3 引力位的球谐展开 ·········· 123
- 5.4 大地水准面 ·········· 124
- 5.5 高程系统 ·········· 125

6 参考椭球和大地坐标系 ·········· 126
- 6.1 旋转椭球 ·········· 126
- 6.2 旋转椭球的曲率及其应用 ·········· 130
- 6.3 大地线 ·········· 132
- 6.4 到参考椭球面的归算 ·········· 133
- 6.5 大地主题解算的正解与反解 ·········· 141

7 地图投影 ·········· 144
- 7.1 地图投影概述 ·········· 144

7.2 投影方程 …………………………………………………………………… 146
7.3 正形投影 …………………………………………………………………… 147
7.4 横墨卡托投影 ……………………………………………………………… 148

8 大地坐标系的建立 …………………………………………………………… 152
8.1 基　准 ……………………………………………………………………… 152
8.2 大地基准 …………………………………………………………………… 154
8.3 坐标变换和基准转换 ……………………………………………………… 155
8.4 天文大地坐标系的定向,最佳吻合椭球 ………………………………… 159
8.5 三维坐标变换 ……………………………………………………………… 160
8.6 WGS-84 …………………………………………………………………… 162
8.7 ITRS 及其实现 …………………………………………………………… 163

1 Introduction

1.1 Definition of Geodesy

Geodesy is the discipline that deals with the measurement and representation (geometry, physics, temporal variations) of the earth and other celestial bodies.

∗∗∗∗∗∗∗∗∗∗∗

According to the classical definition of F. R. Helmert (1880), geodesy is the "science of the measurement and mapping of the earth's surface". This definition has to this day retained its validity; it includes the determination of the earth's external gravity field; as well as the surface of the ocean floor. With this definition, geodesy may be included in the geosciences, and also in the engineering sciences.

∗∗∗∗∗∗∗∗∗∗∗

Webster defines geodesy as "that **branch of applied mathematics** which determines by observation and measurement the exact positions of points and the figures and areas of large portions of the earth's surface, the shape and size of the earth, and the variations of terrestrial gravity." It is a specialized application of several familiar facets of basic mathematical and physical concepts. In practice, geodesy uses the principles of mathematics, astronomy and physics, and applies them within the capabilities of modern engineering and technology.

∗∗∗∗∗∗∗∗∗∗∗

Scientific discipline concerned with the precise figure of the Earth and its determination and significance. Until the advent of satellites, all geodetic work was based on land surveys made by triangulation methods employing a geodesic coordinate system (one used to study the geometry of curved surfaces). It is now possible to use satellites in conjunction with the land-based system to refine knowledge of the earth's shape and dimensions; this endeavour is sometimes termed satellite geodesy.

1.2 Classification of Geodesy

Geodesy may be divided into the areas of global geodesy, control surveying, and plane surveying. **Global geodesy** is responsible for the determination of the figure of the earth including the complete external gravity field. A **control survey** defines the surface of a country by the coordinates of a sufficiently large number of control points. In this fundamental work, the overall curvature of the

---

∗隔开的内容,表示∗前后内容同属该节,但来自不同的文献资料。下同。

earth must be considered. In **plane surveying** (topographic survey, engineer survey, cadastral survey), the details of the land surface are obtained, the horizontal plane is in general sufficient as a reference surface.

There is close interaction between global geodesy, control surveying and plane surveying. The control survey adopts the parameters determined by measurements of the earth, and its own results are available to those who measure the earth. The plane surveys, in turn, are generally tied to the control points of the control surveys and serve then particularly in the development of national map series and in the formation of real estate cadastres.

With the corresponding classifications in the realms of the English and French languages, the concept of "GEODESY" is to be referred only to global geodesy and control surveying.

1.3 Aim of Geodesy

Practical aim of geodesy:

Using the results of scientific geodesy, carry out the measurements and computations needed for making accurate and reliable maps of the earth's surface.

Scientific aim of geodesy:

To determine the size and shape of the earth, and in cooperation with other sciences, study its gravitational field and to some extent the internal structure.

1.4 Historical Introduction

Since about 150 years, geodesy can be regarded as an independent discipline of science. Baeyer's memorandum about the size and figure of the earth "Über die Größe und Figur der Erde" (1861) may be seen as a starting point, even though important geodetic work had been done before by famous scientists such as Newton, Laplace, Gauss, Bessel. But their work was not referred to as geodesy and they did not regard themselves as geodesists. Baeyer's initiative resulted in an extension and unification of existing triangulation and leveling networks covering central Europe. This work was then expanded to the whole of Europe, before its transition to an international effort with the aim to determine the global figure of the earth. It was one of the first international projects in science and the root of what is today the International Association of Geodesy (IAG).

In 2007, 50 years of space age was celebrated. With the launch of Sputnik 1 on October 4, 1957 (and shortly after of Sputnik 2) modern space age began. Already these two satellites had a fundamental effect on geodesy. Almost instantaneously, a large part of 100 years of diligent geodetic work dedicated to the determination of the figure of the earth became out-dated. From measuring the precession of satellite orbits, the earth's flattening could be determined much more accurately than with classical astro-geodetic work. Satellites opened new horizons for geodesy and no other discipline is known to me that has benefited more profoundly from space techniques. Positioning, gravity field determination, earth rotation monitoring and geodetic remote sensing can be done much more accurately, completely and efficiently from space. Geodesy became truly global and three-dimensional. Oceans, a "terra incognita" of the classical times turned with satellites into an area of great geodetic activity. Classical geodetic techniques did not allow the accurate measurement of zenith angles, due to atmospheric refraction. From space, the vertical dimension of the earth's surface can be determined almost as accurately as the horizontal components. Progress of space geodesy was fast and had a great impact. Hand in hand with the rapid development of geodetic space techniques geosciences became more and more interested in geodetic work.

1.5 Recent Developments of Space Geodesy

In recent years, the general emphasis of earth sciences has moved towards **Climate Change and Earth System Science**. Awareness grew that we need a much better understanding of the earth as a system, of solar radiation as its driving force, of the thermal back radiation and how it is affected by even tiny changes in chemical composition of the atmosphere, and last not least, of the impact of man. One fundamental deficiency became particularly evident in the course of the preparation of the last report of the Intergovernmental Panel on Climate Change (Climate Change, 2007) and has been addressed in several articles in Science and Nature: There is a clear lack of observations. Space geodesy is able to provide important new and unique data to Global Change research by measuring mass and energy transport processes in the earth system. Ben Chao (2003) wrote: "After three decades and three orders of magnitude of advances, space geodesy is poised for prime time in observing the integrated mass transports that take place in the earth system, from the high atmosphere to the deep interior of the core. As such space geodesy has become a new remote sensing tool, in monitoring climatic and geophysical changes with ever increasing sensitivity and resolution." One can claim that geodesy, by merging geometry, earth rotation, gravity and geoid, is in a position to provide "metric and weight" to earth system research. Before this background, the establishment of the Global Geodetic Observing System (GGOS) is the right step at the right time. The underlying concept is simple and well described by the scheme shown in Fig. 1.1 and due to Rothacher.

GGOS will combine the **three fundamental pillars of geodesy:** the measurement of the shape of the earth, earth rotation and the earth's gravity field and geoid. The objective is to realize this with a relative precision level of 10^{-9} in one unified earth fixed reference system and to keep this system stable over decades. Where does such a demanding requirement come from? In geosciences, one usually deals with estimates accurate to only a few percent. Global change parameters are small and their temporal changes are slow and even smaller. In general, they cannot be observed directly but have to be derived from a combination of several measurement systems and models. In order to be able to analyze them as a global process they have to be scaled relative to the dimension of the earth. Let us take an example. Sea level at an arbitrary tide gauge may vary by a few meters, due to tides and storm surges. Measurement of sea level change with a precision of a few mm requires therefore a relative precision of 10^{-3} at this particular station. Local sea level monitoring can be transformed into a global monitoring system by satellite systems such as altimetry and GPS. Only then, a global process can be deduced from local tide gauge records. In order to achieve cm-or mm-precision with satellite systems globally, orbit determination and altimetric measurements have to be delivered with a relative precision of 1 ppb.

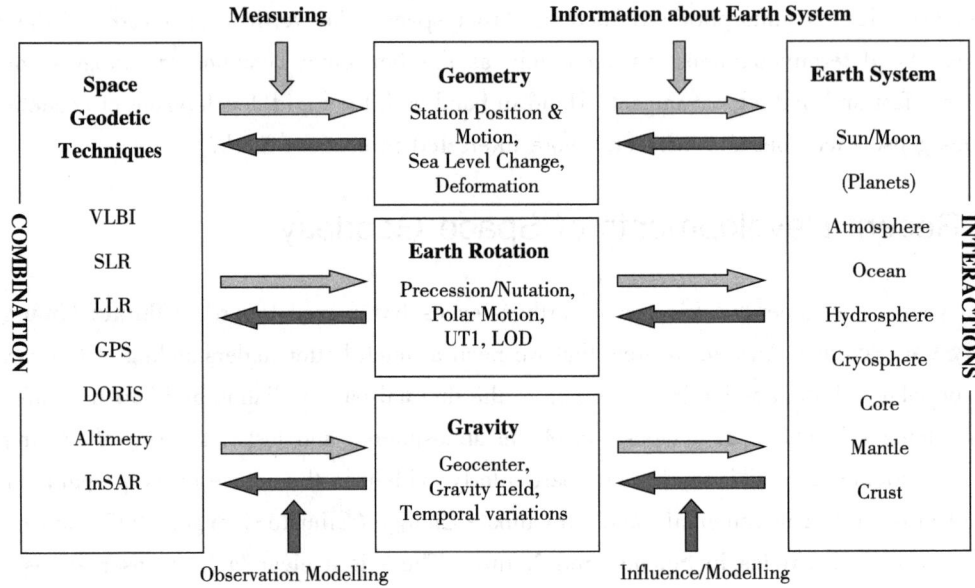

Fig. 1.1　Combination of the space geodetic techniques, the three pillars of space geodesy, and the interaction with the components of the earth's system

In order to meet the goals set for GGOS a series of rather fundamental geodetic problems have to be dealt with. The three pillars of geodesy, geometry, earth rotation and gravity have to be expressed in one and the same earth fixed reference system with millimeter precision and stability (of the frame) has to be guaranteed over decades. This requires the space as well as the ground segments to function as one homogeneous entity as if all observations were done in one observatory encompassing the earth. Each observation contains a superposition of a variety of effects, related to ionosphere, atmosphere, oceans, ice shields and solid earth. In order to employ them for earth system research strategies have to be developed for their separation and quantification by analyzing their spatial, temporal and spectral characteristics. Satellite measurements represent time series along their orbit. Via the earth's rotation and the choice of the satellite orbit elements these time series are related to a spatial and temporal sampling of the earth. The reconstruction of the temporal and spatial geophysical phenomena poses a complicated problem of aliasing and inversion. The current investigations of the global water cycle or of the ice mass balance in Greenland and Antarctica from GRACE gravimetry are exactly problems of this type. The inclusion of terrestrial and airborne data, such as surface loading, ocean bottom pressure, tide gauges, gravimetry or altimetry may certainly help. However, this step is not easy either, because terrestrial measurements are affected by local influences and exhibit a spectral sensitivity quite different from that of satellite observations. Probably the most effective support to de-aliasing and separation of geophysical phenomena is the inclusion of prior information, such as models of solid earth and ocean tides, atmosphere, oceans, ice, hydrology or glacial isostatic adjustment, however only if they are introduced consistently for all techniques of the observing system. Important work towards these goals is currently underway and we see geodetic techniques used much more widely in the various earth disciplines.

Wegener developed over the years from an almost mono-disciplinary project to regional multi-disciplinary activity combining a large variety of geodetic and non-geodetic measurements techniques and involving all geo-disciplines relevant to its objectives. It is therefore an excellent example on how GGOS could operate on a global scale.

Vocabulary

altimetry /æl'timitri/ *n.* 测高学,高度测量法(以海平面为基准)
celestial /sə'lestʃəl/ *adj.* 天的,天空的
curvature /'kə:vəˌtʃʊə, -tʃə/ *n.* 曲率,弯曲
earth rotation /ə:θ rəu'teiʃən/ 地球自转
figure of the earth /'figə ɔv ði ə:θ/ 地球形状
flattening /'flætniŋ/ *n.* 扁率
geodesy /dʒi:'ɔdisi/ *n.* 大地测量学
geoid /'dʒi:ɔid/ *n.* 大地水准面
gravity field /'græviti fi:ld/ 重力场
leveling /'levəliŋ/ *n.* 水准测量
leveling network /'levəliŋ 'netwə:k/ 水准网
network /'netwə:k/ *n.* (控制)网
precession /pri'seʃən/ *n.* 岁差,进动
triangulation /traiˌæŋgju'leiʃən/ *n.* 三角测量
zenith /'zi:niθ/ *n.* 天顶

2 Reference System

2.1 Time Systems

In the context of this book, time is regarded as absolute and independent of space. In reality, time keeping, and satellite measurements in general, are so accurate nowadays that adequate modeling requires the use of special and general theory of relativity for practical reason.

Time keeping requires a **periodic process**, a **counter** (in order to count the number of periods) and an **origin** where the counting starts. In addition, in order to be able to keep the same time at different locations **some means of transfer/transport of time** has to be available. There exist a number of natural "clocks" that produce very stable periodic oscillations: the orbit of the earth about the sun, of the moon about the earth and earth rotation. Their fundamental periods, year, month and day, are closely related to natural processes such as seasons that affect our living conditions and these periods define the basic structure of our life. From these fundamental periods the basic long term counting structure has been deduced, our **calendars** (we use the Gregorian calendar, adopted in 1582). In scientific work a continuous counting is preferable to the complicated structure of counting with months or year of varying length. For this purpose the **Julian date (JD)** has been invented with 36525 days per century.

The adopted reference date is

$$J2000.0 = 2000 \text{ Jan } 1.5 = \text{January } 1, 2000 \text{ at } 12h$$

where it is

$$JD\ 2451545.0.$$

For a long time the natural period day, and even more the revolution of the moon, were superior in terms of stability to any artificial clock. Only with the advent of quartz and atomic oscillators, artificial clocks were created that meanwhile surpassed the precision and stability of natural clocks. Our current definition of the unit of second is based on the oscillation period of a cesium clock. In 1984 **atomic time**(Temps Atomique International = **TAI**) has been introduced as official, internationally adopted time. It has a constant off-set of 32s.184 with respect to the **terrestrial dynamic time (TDT)**. The latter is derived from models of planetary motion and based on the theory of relativity. TAI has a constant off-set of 19s with respect to **GPS-time**.

Civilian time is related to the rhythm of day and night, i.e. to the rise and fall of the sun. Because of the complicated deviations of the apparent motion of the sun, some model or mean solar motion has been conceived. It refers to the Greenwich meridian and is denoted **universal time (UT)**. From UT standard zonal times have been deduced. earth rotation—and therefore UT—exhibits a drift and small irregular fluctuations (changes in LOD) with respect to TAI. In order to circumvent this, a coordinated universal time (**UTC**) has been conceived, which on the one hand is kept synchronous with respect to TAI and on the other hand, through regular corrections (leap seconds), is kept within small bounds to follow the actual angular rate of the earth. The actual and uncorrected universal time is denoted **UT1**. It represents the actual phase angle of the rotating earth. The difference UT1-UTC is provided in monthly tables and provided as coded message in broadcasted time signals. If the difference UT1-UTC exceeds the size of 0.9 s a leap second is introduced. **UT0** completes the system of universal times. It contains all variations in rotation due to polar motion.

Finally, **sidereal time** is the angle of a terrestrial meridian (rotating with the earth) with respect to vernal equinox. The most prominent types of sidereal time are Greenwich Mean Sidereal Time (GMST) $\bar{\Theta}$ and Greenwich Apparent (or true) Sidereal Time (GAST) Θ.

GMST is corrected for fluctuations caused by nutation. It is

$$\Theta = \bar{\Theta} + \Delta n = \bar{\Theta} + \Delta\psi \cdot \cos\varepsilon + 0.00264'' \cdot \sin\Omega + 0.000063'' \cdot \sin2\Omega \qquad (2.1)$$

and Ω the mean node of the moon. Greenwich Apparent Sidereal Time is needed for the transformation from earth-fixed to space-fixed.

Since sidereal time is measured with respect to vernal equinox, while universal time is a solar time and counted with respect to the apparent pass of the sun through the meridian at Greenwich the length of the year is different by one day: tropical year

in solar days: 365.24220

in sidereal days: 366.24220.

This difference has to be accounted for when transforming UT1 to GMST. It holds

$$\text{GMST} = \text{UT1} + \alpha(\Theta) - 12^h \qquad (2.2)$$

with the right ascension of the sun:

$$\alpha(\Theta) = 12^h + (24110.54840^s + 8640.812866^s \cdot t + 0.093104^s \cdot t^2 - 6.2^s \cdot 10^{-6} \cdot t^3) \qquad (2.3)$$

and

$$t = (T - \text{J2000.0})/36525.0 \qquad (2.4)$$

with T Julian Date at epoch and J2000.0 = JD2451545.0.

2.2 International Earth Rotation Service

The International Earth Rotation Service (IERS) is in charge of **providing and maintaining conventional celestial and terrestrial reference frames**. These frames are a realization of the reference systems recommended by the International Astronomical Union (IAU) and the International Union of Geodesy and Geophysics (IUGG). IERS is also responsible for **the determination of the orientation parameters of the earth as functions of time**, which relate the two frames to each other.

Established by the IAU and IUGG, the IERS has operated since January 1, 1988. It collects, analyzes, and models observations of a global network of astronomic and geodetic stations (about 300 sites in 1996), operating either permanently or for a certain time span. Observation techniques include VLBI, LLR, SLR, GPS, DORIS.

The different types of observations are evaluated at the respective IERS coordinating centers and then combined by an adjustment at the IERS Central Bureau. The results include the positions (coordinates) of both the extragalactic radio sources and the terrestrial stations, the earth orientation parameters (EOP), and other information. With respect to the EOP, VLBI provides information about precession, nutation, polar motion, and UT1. Satellite techniques contribute to the daily interpolation of UT and to the determination of polar motion. The results are disseminated through bulletins, annual reports, and technical notes. The evaluation of the observations is based on the IERS Conventions, which are consistent with the IAU and IUGG/IAG recommendations for reference systems.

2.3 Celestial Reference System

An inertial system is needed in order to describe the motions of the earth and other celestial bodies in space, including those of artificial satellites. Such a system is characterized by Newton's law of motion; it is either at rest or in the state of a uniform rectilinear motion without rotation. A space-fixed system (celestial reference system) represents an approximation to an inertial system and can be defined by appropriate conventions: Conventional Inertial System (CIS). The coordinate frame for such a system is provided by spherical astronomy. The spatial orientation of this frame varies with time, and therefore, modeling of the variations is required. The International Celestial Reference Frame represents the realization of the celestial reference system.

2.3.1 Equatorial System of Spherical Astronomy

The coordinates of the celestial reference system are **defined by the equatorial system of spherical astronomy**. We introduce a three-dimensional Cartesian coordinate system with the origin at the center of mass of the earth (geocenter). The Z-axis coincides with the rotational axis of the earth. The X and Y-axes span the equatorial plane, with the X-axis pointing to the vernal equinox and the Y-axis forming a right-handed system (Fig. 2.1).

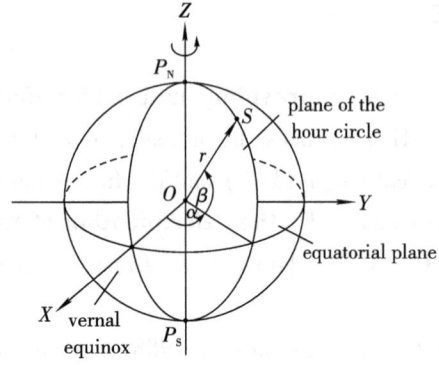

Fig. 2.1 Astronomic equatorial system

We circumscribe the unit sphere (celestial sphere) about the earth. The rotational axis meets the sphere at the celestial north and south poles P_N and P_S. The great circles perpendicular to the celestial equator, which contain the celestial poles, are called **hour circles**, and the small circles parallel to the equator are termed **celestial parallels**.

The **right ascension** α is the angle measured in the plane of the equator between the planes of the hour circles passing through the vernal equinox and the celestial body S; it is reckoned from the vernal equinox anticlockwise. The **declination** δ is the angle measured in the plane of the hour circle between the equatorial plane and the line OS (positive from the equator to P_N and negative to P_S).

The position of a celestial body S can be described either by the Cartesian coordinates X, Y, Z, or by the spherical coordinates α, δ, r (r = distance from the origin O). We have the transformation

$$\boldsymbol{r} = \begin{pmatrix} X \\ Y \\ Z \end{pmatrix} = r \begin{pmatrix} \cos\alpha\cos\delta \\ \sin\alpha\cos\delta \\ \sin\delta \end{pmatrix} \tag{2.5}$$

In geodesy, only directions are important for stars and extragalactic sources. With $r = 1$, α and δ describe the position of S on the unit sphere. They can also be expressed by the lengths of the corresponding arcs on the equator and the hour circle.

We introduce the local meridian plane of the observer, spanned by the local vertical (direction of the plumb line) and the rotational axis, after a parallel shift from the geocenter to the topocenter. The zenith point Z is the intersection of the vertical with the unit sphere, and the celestial meridian is the great circle through Z and the poles (Fig. 2.2). The

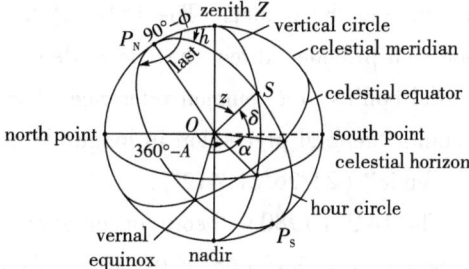

Fig. 2.2 Astronomic equatorial and horizon system

hour angle h is measured in the equatorial plane between the celestial meridian through Z and the hour circle of S, reckoned from the upper meridian toward west. Because of the earth's rotation, the hour angle system (h, δ) depends on time. The h, δ-system is rotated, with respect to the α, δ-system, about the polar axis by the angle of sidereal time LAST. We have the relation

$$\text{LAST} = h + \alpha \tag{2.6}$$

which is used with time determination.

2.3.2 Precession and Nutation

The earth's axis of rotation, which has been introduced as the Z-axis, changes its spatial orientation with time. As a consequence, the position (α, δ) of a celestial body varies, with a superposition of long and short-periodic effects.

The **lunisolar precession** is a long-periodic effect caused by the gravitation of the moon and the sun on the equatorial bulge of the earth. This creates a force couple (torque) which tends to turn the equatorial plane into the plane of the ecliptic (Fig. 2.3). In combination with the moment of the earth's rotation, the earth's axis describes a gyration of a cone with a generating angle of 23.5° (corresponding to the obliquity of the ecliptic ε), about the northern pole of the ecliptic E_N. The vernal equinox moves clockwise along the ecliptic at a rate of 50.3″/year, making a complete revolution in about 25 800 years. The gravitation of the planets causes a slow dislocation of the earth's orbit and thereby an additional migration of the vernal equinox along the equator and a change in ε: **planetary precession**. The sum of the lunisolar and the planetary precession is termed general precession.

The precession is superimposed by short-periodic effects known as **nutation**, which has periods between 5 days and 18.6 years. These periods are mainly due to the time variations of the inclination of the moon's orbit with respect to the ecliptic (appr. 5°). Other components have semiannual and semimonthly periods and stem from the oscillations of the sun and moon between the earth's northern and southern hemisphere.

Precession and nutation can be modeled as a function of time using the ephemeredes of the moon, the sun, and the planets. The IAU (1976) theory of precession provides three time-dependent Eulerian rotation angles for reducing the positions of celestial bodies to a common reference. For the reference epoch J2000.0, we have the fundamental constants "general precession in longitude at the ecliptic" (5 029.096 5″/century) and "obliquity of the ecliptic" (23°26′21.412″).

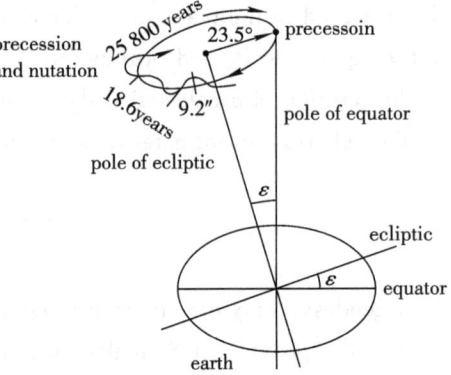

Fig. 2.3 Precession and nutation

The IAU (1980) theory of nutation describes this effect by a rotation about the cone of precession. The deviation of the true pole from the mean pole is modeled by two time-dependent parameters. Hereby, the earth is regarded as an elliptical, rotating, elastic, and ocean-free body with solid inner and liquid outer cores. For the epoch J2000.0 the constant of nutation is 9.202 5″.

The IAU models for precession and nutation define the reference pole for the international celestial reference frame (Celestial Ephemeris Pole CEP). CEP is free of diurnal or quasidiurnal nutation terms (amplitudes <0.001″) with respect to the space- or earth-fixed coordinate systems. It is also referred to as the pole of the instantaneous equatorial system.

The IAU models for precession and nutation provide a precision of ±0.001″ at 5 to 7 days resolution. An improved theory has been developed at the IERS based on recent VLBI and LLR data. Larger offsets (<0.02″) of the celestial pole from CEP have been found; these are published regularly by IERS.

The instantaneous position of a celestial body is called true position at the epoch t. By accounting for nutation, we obtain the mean position at epoch t, which refers to the mean celestial equator and the mean vernal equinox. If precession is also taken into account, we get the mean position at the reference epoch J2000.0.

2.3.3 International Celestial Reference Frame

The International Celestial Reference System (ICRS), as recommended by IAU, is based on the general theory of relativity, with the time coordinate defined by the international atomic time. **ICRS approximates a space-fixed conventional inertial-system (CIS) with the origin at the barycenter of the solar system.** It is assumed that no global rotation of the system exists. This implies that the defining sources are either free from proper motion (component of spatial motion tangent to the celestial sphere) or that this motion can be modeled. The coordinate axes are defined by the celestial reference pole and the vernal equinox as provided by the IAU models for precession and nutation. They are **realized through mean directions to extraterrestrial fiducial objects:** stellar or radio source CIS.

The stellar system is based on the stars of the Fundamental Catalogue FK5. It provides the mean positions (α, δ) and the proper motions (generally $<1''$/year) of 1 535 fundamental stars for the epoch J2000. 0, with precisions of $\pm 0.01...0.03''$ and $\pm 0.05''$/century respectively. A supplement to FK5 contains additional stars up to an apparent magnitude of 9.5. The mean equator and the mean vernal equinox for J2000. 0 are realized by the FK5 catalogue, with an accuracy of $\pm 0.05''$. Due to refraction uncertainties, earth-based astrometry can hardly improve this accuracy.

Astronomic space missions have significantly improved the realization of a stellar CIS. The HIPPARCOS astrometry satellite (ESA, 1989—1993) was used to construct a network by measuring large angles between about 100 000 stars (up to an apparent magnitude of 9) covering the entire sky. The reference frame thus established provides an accuracy of $\pm 0.001''$ and $\pm 0.0005''$/year for proper motion. From improved FK5 data and Hipparcos results, an FK6 catalogue has been developed for a small number of stars (340 "astrometrically excellent"), resulting in an improvement of proper motion as compared to the Hipparcos catalogue. Future astrometric space missions will employ optical interferometry and thus increase the positional accuracy to $\pm 0.00001''$.

The radio source system is based on extragalactic radio sources (quasars and other compact sources). It was adopted as ICRS by the IAU in 1997 and has superseded the previous stellar system (FK5) since 1998. Due to the large distances (>1.5 billion light years), these sources do not show a measurable proper motion. The system is realized through the International Celestial Reference Frame (ICRF), established and maintained by IERS. ICRF contains the coordinates (equatorial system, epoch J2000.0) of more than 600 objects. About 200 of them are well observed "**defining sources**", and 100 more are used for densification and connection to the stellar-fixed reference system. The southern sky is not as well covered, as the telescopes are concentrated in the northern hemisphere. The coordinates of the radio sources are determined by radio astronomy, with a precision of better than $\pm 0.001''$ on the average and $\pm 0.0003''$ for the most precisely observed objects.

2.4 Terrestrial Reference System

An earth-fixed reference system is introduced **for positioning and navigation on and close to the earth's surface and for describing the earth's gravity field as well as other physical parameters**. It is defined by a three-dimensional geocentric coordinate system. The orientation of this system changes with time and with respect to the solid earth's body as well as to the celestial reference system. The system is realized by the IERS International Terrestrial Reference Frame.

2.4.1 Global Earth-Fixed Geocentric System

An earth-fixed (i.e., rotating with the earth) system of spatial Cartesian-coordinates X, Y, Z is used as the fundamental terrestrial coordinate system. Its **origin** is at the earth's center of mass (geocenter), being defined for the whole earth including hydrosphere and atmosphere. The **Z-axis** is directed towards a conventional "mean" terrestrial (north) pole. The "mean" equatorial plane is perpendicular to it and contains the **X and Y-axes**. A "mean" rotational axis and equatorial plane has to be introduced because the rotation of the earth changes with respect to the earth's body over time. The XZ-plane is generated by the conventional "mean" meridian plane of Greenwich, which is spanned by the mean axis of rotation and the Greenwich zero meridian, to which Universal Time refers. The Z and X-axes are realized indirectly through the coordinates of terrestrial "fiducial" stations. The Y-axis is directed so as to obtain a right-handed system.

2.4.2 Polar Motion, Length of Day, Geocenter Variations

The rotation of the earth can be described by a vector directed to the north pole of the instantaneous axis of rotation and by the angular velocity ω.

Direction and magnitude of the rotational vector change with time due to astronomical and geophysical processes. These processes include variations of the lunar and solar gravitation and mass redistributions in the atmosphere, the hydrosphere, the solid earth, and the liquid core. The changes are secular, periodic or quasiperiodic, and irregular in nature.

Polar motion (wobble) is the motion of the rotation axis relative to the earth's crust as viewed from the earth-fixed reference system. It directly affects the coordinates of stations on the earth's surface and the gravity vector. Polar motion consists of several components:

(1) A free oscillation with a period of about 435 days (Chandler period), with an amplitude of $0.1''$ to $0.2''$, in a counterclockwise sense as viewed from the north pole. The Chandler wobble is due to the fact that the spin axis of the earth does not coincide exactly with a principal axis of inertia.

(2) The Chandler wobble is superposed by an annual oscillation forced by seasonal displacements of air and water masses. It proceeds in the same direction as the Chandler wobble with amplitudes of $0.05''$ to $0.1''$.

(3) A secular motion of the pole has been observed for more than 100 years. The motion consists of an irregular drift of about $0.003''$/year in the direction of the $80°W$ meridian. Secular motion is mainly due to the melting of the polar ice and to large-scale tectonic movements; it attains large amounts over geological epochs: polar wander.

(4) More irregular variations occur at time scales from a few days to years with amplitudes up to $0.02''$. They originate primarily from mass redistributions within the atmosphere, but variations due to ocean volume changes, ground water variations, and earthquakes also occur.

The superposition of these components results in a slightly perturbed spiral line curve of the instantaneous pole with a slowly advancing mean position. Over one year, the deviations from the mean position remain $<0.3''$, corresponding to 9 m on the earth's surface.

The reference for describing the actual position of the pole with respect to the solid earth is provided by the IERS reference pole. It agrees within $\pm 0.03''$ with the Conventional International Origin (CIO), which was defined by the mean position of the north pole as determined between 1900.0 and 1906.0. The position of the instantaneous pole (Celestial Ephemeris Pole) with respect to the reference pole is given by the rectangular coordinates x_P, y_P, which are defined in the plane tangential to the pole. The x_P-axis is in the direction of the Greenwich mean meridian (consistent with the previous BIH zero meridian), and the y_P-axis is directed along the 90°W meridian. These plane coordinates are usually expressed as spherical distances (in units of arcsec) on the unit sphere.

The angular velocity ω of the earth's rotation, as monitored from the earth, changes with time. Relative changes may reach several 10^{-8}, which corresponds to several ms for one day. The variations are generally described by the excess revolution time with respect to 86 400 s and then called **Length Of Day (LOD)**. They are derived by comparing astronomical time determinations, which deliver Universal Time UT1, with the uniform time scales TAI or UTC.

The following components of LOD variations have been observed:

(1) A secular decrease in the angular velocity of the earth's rotation is caused mainly by tidal friction. It lengthens the day by about 2ms/century.

(2) Fluctuations over decades are due to motions in the earth's liquid core and to slow climatic variations.

(3) The tides of the solid earth and the oceans produce variations of about 1ms with long (annually) and short (monthly and less) periodic parts.

(4) Seasonal effects are explained by atmospheric excitation, with contributions from water and ice budget variations.

(5) More irregular oscillations stem from different sources, such as terrestrial mass displacements (earthquakes), solar activity, and atmospheric events, e.g. El Nino.

While the effect of polar motion on observations is dependent on location, LOD changes act uniformly on all points. The pole coordinates and LOD, as well as ω, are provided as Earth Orientation Parameters (EOP) by the IERS with daily resolution and accuracy of $\pm 0.0003''$ resp. ± 0.02 ms or better.

The position of the **geocenter** (origin of the terrestrial reference system) changes slightly in time with respect to the monitoring observatories. Annual and semiannual variations have been found, with amplitudes of several mm/year, from the analysis of satellite orbits. The variations are caused primarily by mass redistributions in the atmosphere and the oceans and by continental water variations. Through the coordinates of the ITRF stations, the geocenter is given with an accuracy of a few mm.

2.4.3 International Terrestrial Reference Frame

The International Terrestrial Reference System is realized by the IERS through a global set of space geodetic observing sites. The geocentric Cartesian coordinates and velocities of the observing sites comprise the International Terrestrial Reference Frame (ITRF). The stations participating in the ITRF carry out observations either continuously or at certain time intervals. Observations are made on twelve of the larger tectonic plates, which permits the derivation to station velocities related to plate tectonics.

Annual realizations of the ITRF are published by the IERS. The ITRF 97 is comprised of the geocentric positions (X, Y, Z) for more than 550 stations at about 320 sites and corresponding site velocities. The accuracy of the results depends on the observation techniques and is maximum for VLBI, SLR, and GPS observations. Several time variable effects are taken into account, including displacements due to the solid earth tides, ocean and atmospheric loading effects, and postglacial rebound. The ITRF solutions satisfy the condition of no residual net-rotation relative to the plate tectonics model NNR-NUVEL1A; vertical movements are not allowed at all. The orientation of the ITRF is given with respect to the IERS reference pole and reference meridian. The actual (time t) position vector r of a point on the earth's surface is derived from its position at the reference epoch (t_0) by

$$r(t) = r_0 + \dot{r}_0(t - t_0) \tag{2.7}$$

where r_0 and $\dot{r}_0$ are the position and velocity respectively at t_0.

2.5 Gravity Field Related Reference Systems

Most geodetic and astronomic observations on or close to the earth's surface refer to the earth's gravity field by orientation along the local vertical. Consequently, local gravity-field-related reference systems are introduced for the modeling of these observations. The orientation of the local systems with respect to the global reference system is given by astronomic latitude and longitude. These orientation parameters are used for transformation from the local systems into the global system and back.

2.5.1 Orientation of the Local Vertical

The direction of the plumb line (local vertical) with respect to the global geocentric system is given by two angles (Fig. 2.4). **The astronomic (geographic) latitude** φ is the angle measured in the plane of the meridian between the equatorial plane and the local vertical through the point P. It is reckoned positive from the equator northward and negative to the south. The angle measured in the equatorial plane between the Greenwich meridian plane and the plane of the meridian passing through P is the **astronomic (geographic) longitude** λ; it is reckoned positive toward the east. The gravity potential W locates P in the system of level surfaces W = const.. The local astronomic meridian plane is spanned by the local vertical at P and a line parallel to the rotational axis.

We introduce a system of non-linear "natural" coordinates φ, λ, W defined in the gravity field. Astronomical latitude φ and astronomical longitude λ describe the direction of the plumb line at the point P. The gravity potential W locates P in the system of level surfaces W = const.. Hence, P is determined by the non-orthogonal intersection of the coordinate surfaces φ = const., λ = const. and the surface W = const.. The coordinate lines (spatial curves) are called astronomic meridian curve (λ, W = const.), astronomic parallel curve (φ, W = const.), and isozenithal line (φ, λ = const.).

The natural coordinates can be determined by measurements. Astronomic positioning provides latitude and longitude. Although W cannot be measured directly, potential differences can be derived from leveling and gravity measurements and then referred to a selected level surface, e.g., the geoid.

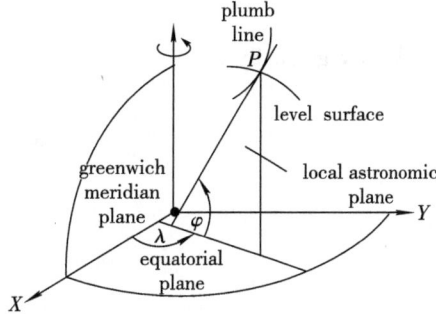

Fig. 2.4 Astronomic latitude and longitude

2.5.2 Local Astronomical Systems

Geodetic and astronomic observations are tied to the direction of the plumb line at the point of observation and thereby to the earth's gravity field. An exception is distance measurements, which are independent of the reference system. Thus, these observations establish local gravity-field related systems: Local astronomical systems. Their origin is at the point of observation P. The Z-axis coincides with the local vertical and points toward the zenith. The X-axis (north) and the Y-axis (east) span the horizontal plane, which is tangent to the level surface $W = W_P$. This x, y, z-system is left-handed.

2.5.3 Transformation between the local astronomic and the geocentric system

The plumb line can be referred to the global geocentric-system by means of the "orientation" parameters astronomic latitude φ and longitude λ (Fig. 2.5). After a parallel shift of the global system into the local one, we transform the latter one to a right-handed system by applying the **reflection matrix**

$$S_2 = \begin{pmatrix} 1 & 0 & 0 \\ 0 & -1 & 0 \\ 0 & 0 & 1 \end{pmatrix}$$

We then rotate the local system by $90° - \varphi$ around the (new) y-axis and by $180° - \lambda$ around the z-axis with the **rotation matrices**

$$R_2(90° - \varphi) = \begin{pmatrix} \sin\varphi & 0 & -\cos\varphi \\ 0 & 1 & 0 \\ \cos\varphi & 0 & \sin\varphi \end{pmatrix}$$

$$R_3(180° - \lambda) = \begin{pmatrix} -\cos\lambda & \sin\lambda & 0 \\ -\sin\lambda & -\cos\lambda & 0 \\ 0 & 0 & 1 \end{pmatrix}$$

Coordinates differences between P_i and P in the geocentric system are thus obtained by

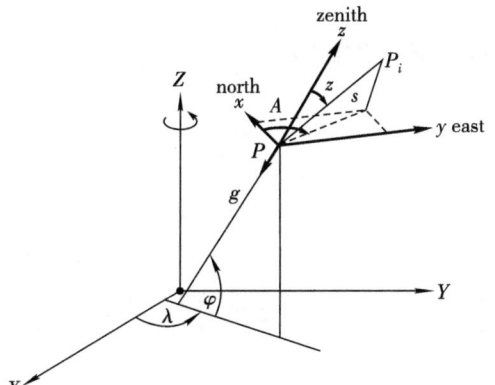

Fig. 2.5 Local astronomic and global geocentric system

$$\Delta X = Ax \tag{2.8}$$

With x given by

$$x = \begin{pmatrix} x \\ y \\ z \end{pmatrix} = s \begin{pmatrix} \cos A \sin z \\ \sin A \sin z \\ \cos z \end{pmatrix}$$

and

$$\Delta X = \begin{pmatrix} \Delta X \\ \Delta Y \\ \Delta Z \end{pmatrix}$$

The transformation matrix reads as

$$A = R_3(180° - \lambda) R_2(90° - \varphi) S_2 = \begin{pmatrix} -\sin\varphi\cos\lambda & -\sin\lambda & \cos\varphi\cos\lambda \\ -\sin\varphi\sin\lambda & \cos\lambda & \cos\varphi\sin\lambda \\ \cos\varphi & 0 & \sin\varphi \end{pmatrix} \tag{2.9}$$

The inversion of (2.8) is performed easily considering that A is orthonormal:

$$A^{-1} = A^{T}$$

We obtain

$$x = A^{-1} \Delta X \tag{2.10}$$

with

$$A^{-1} = \begin{pmatrix} -\sin\varphi\cos\lambda & -\sin\varphi\sin\lambda & \cos\varphi \\ -\sin\lambda & \cos\lambda & 0 \\ \cos\varphi\cos\lambda & \cos\varphi\sin\lambda & \sin\varphi \end{pmatrix} \tag{2.11}$$

Equations (2.8) to (2.11) are the basic equations for the evaluation of local geodetic measurements within the three-dimensional reference frame.

Vocabulary

astronomical system /ˌæstrɔnəmik ˈsistəm/ 天文系统
Cartesian coordinate system /kɑːˈtiːzjən kəuˈɔːdineit ˈsistəm/ 笛卡儿坐标系(直角坐标系)
Conventional International Origin /kənˈvenʃnəl ˌintəˈnæʃənəl ˈɔridʒin/ 国际协议原点
distance measurements /ˈdistəns ˈmeʒəmənt/ 距离测量
earth orientation parameters (EOP) /əːθ ˌɔːrienˈteiʃən pəˈræmitəs/ 地球定向参数
equatorial /ˌiːkwəˈtɔːriːəl/ adj. 赤道的
Fundamental Catalogue /ˌfʌndəˈmentəl ˈkætəlɔg/ 基本星表
geocenter /ˌdʒiːəuˈsentə/ n. 地球质量中心
geocentric system /ˌdʒiːəuˈsentrik ˈsistəm/ 地心体系
Greenwich meridian /ˈgrinidʒ məˈridiːən/ 格林尼治子午线
interferometry /ˌintəfiəˈrɔmitri/ n. 干涉测量法
latitude /ˈlætitjuːd/ n. 纬度
level surface /ˈlevl ˈsəːfis/ 水准面
longitude /ˈlɔndʒitjuːd/ n. 经度
meridian /məˈridiːən/ n. 子午圈,子午线

nutation /njuːˈteiʃən/ n. 章动
orientation /ˌɔːrienˈteiʃən/ n. 方向, 定向
plate tectonics /pleit tekˈtɔniks/ 板块构造学
plumb line /plʌm lain/ 铅垂线
polar motion /ˈpəulə ˈməuʃən/ 极移
sidereal time /saiˈdiəriːəl taim/ 恒星时
terrestrial reference frame /təˈrestriːəl ˈrefrəns freim/ 地球参考框架
universal time /ˌjuːniˈvəːsəl taim/ 世界时
vernal equinox /ˈvəːnəl ˈiːkwinɔks/ 春分点

3 Methods of Measurement in Geodesy

3.1 Terrestrial Geodetic Measurements

Terrestrial geodetic measurements are carried out by directly observing geometric quantities between points on the earth's surface. The majority of the observations refer to the gravity field of the earth through orientation in the local astronomical systems. The measurement of horizontal and zenith angles and distances allow relative positioning, where combined instruments (total stations) are generally used. Very precise height differences are provided by leveling.

Due to the high accuracy and economy of satellite-based measurement techniques, terrestrial geodetic measurements are used primarily for interpolating satellite-derived results or in areas where satellite methods fail or need terrestrial support (underground and underwater positioning, forests, urban areas, engineering surveys, local geodynamics).

3.1.1 Horizontal Angle Measurements

Triangulation

In navigation, surveying, and civil engineering, a technique for precise determination of a ship's or aircraft's position, and the direction of roads, tunnels, or other structures under construction. It is based on the laws of plane trigonometry, which state that, if one side and two angles of a triangle are known, the other two sides and angle can be readily calculated. One side of the selected triangle is measured; this is the baseline. The two adjacent angles are measured by means of a surveying device known as a theodolite, and the entire triangle is established. By constructing a series of such triangles, each adjacent to at least one other triangle, values can be obtained for distances and angles not otherwise measurable. Triangulation was used by the ancient Egyptians, Greeks, and other peoples at a very early date, with crude sighting devices that were improved into the diopter, or dioptra (an early theodolite), and were described in the 1st century AD by Heron of Alexandria.

Theodolite

Basic surveying instrument of unknown origin but going back to the 16th-century English mathematician Leonard Digges; it is used to measure horizontal and vertical angles. In its modern form it consists of a telescope mounted to swivel both horizontally and vertically. Leveling is accomplished with the aid of a spirit level; crosshairs in the telescope permit accurate alignment with the object sighted. After the telescope is adjusted precisely, the two accompanying scales, vertical and horizontal, are read.

Mounted on a tripod with adjustable legs, the theodolite is used in the field to obtain precise angular measurements for triangulation in road building, tunnel alignment, and other civil-engineering work. The transit is a variety of theodolite that has the telescope so mounted that it can be completely reversed, or transited. The phototheodolite, a combination camera and theodolite mounted on the same tripod, is used in terrestrial photogrammetry for mapmaking and other purposes.

3.1.2 Distance Measurements

Trilateration

Method of surveying in which the lengths of the sides of a triangle are measured, usually by electronic means, and, from this information, angles are computed. By constructing a series of triangles adjacent to one another, a surveyor can obtain other distances and angles that would not otherwise be measurable. Formerly, trilateration was little used in comparison to triangulation, a method for determining two sides and an angle of a triangle from the length of one side and two angles, because of the difficulty of the computations involved. But the development of electronic distance-measuring devices has made trilateration a common and preferred system. Except that only lines are measured, while all angles are computed, the field procedures for trilateration are like those for triangulation.

EDM

Distance measurement was impractical over large or hilly areas until the invention of electromagnetic distance measurement (EDM) in the mid-20th century. This procedure has made it possible to measure distances as accurately and easily as angles, by electronically timing the passage of radiation over the distance to be measured; microwaves, which penetrate atmospheric haze, are used for long distances and light or infrared radiation for short ones. In the devices used for EDM, the radiation is either light (generated by a laser or an electric lamp) or an ultrahigh-frequency radio beam. The light beam requires a clear line of sight; the radio beam can penetrate fog, haze, heavy rain, dust, sandstorms, and some foliage. Both types have a transmitter-receiver at one survey station. At the remote station the light type contains a set of corner mirrors; the high-frequency type incorporates a retransmitter (requiring an operator) identical to the transmitter-receiver at the original station. A corner mirror has the shape of the inside of a corner of a cube; it returns light toward the source from whatever angle it is received, within reasonable limits. A retransmitter must be aimed at the transmitter-receiver.

In both types of instrument, the distance is determined by the length of time it takes the radio or light beam to travel to the target and back. The elapsed time is determined by the shift in phase of a modulating signal superimposed on the carrier beam. Electronic circuitry detects this phase shift and converts it to units of time; the use of more than one modulating frequency eliminates ambiguities that could arise if only a single frequency had been employed.

EDM instrument

The electromagnetic distance measurement (EDM) instruments are the geodimeter and the tellurometer. These instruments measure slope distance, which may be reduced to ellipsoid with adequate accuracy by means of geodetic height.

3.1.3 Trigonometrically Determined Heights

For faster work in hilly areas, where lower accuracies usually are acceptable, trigonometric height determination is employed using a theodolite to measure vertical angles and measuring or calculating the distances by triangulation. This procedure is particularly useful in obtaining heights throughout a major framework of triangulation or traverse where most of the points are on hilltops. To increase precision, the observations are made simultaneously in both directions so that aerial refraction is eliminated; this is done preferably around noon, when the air is well mixed.

3.1.4 Leveling

In spirit leveling the surveyor has for centuries used a surveying level, which consists of a horizontal telescope fitted with cross hairs, rotating around a vertical axis on a tripod, with a very sensitive spirit level fixed to it; the instrument is adjusted until the bubble is exactly centered. The reading on a graduated vertical staff is observed through the telescope. If such staffs are placed on successive ground points, and the telescope is truly level, the difference between the readings at the cross hairs will equal that between the heights of the points. By moving the level and the staffs alternately along a path or road and repeating this procedure, differences in height can be accurately measured over long horizontal distances.

3.2 Satellite Observations

Satellite geodesy utilizes artificial satellites and the moon as extraterrestrial targets or sensors. Classical measurement methods, introduced and employed from the 1960's to the 1980's, demonstrated the efficiency of satellite observations for establishing large-region geodetic control networks and gravitational field determination. Today, the Global Positioning System (GPS) governs three-dimensional positioning at all scales, while laser distance-measurements primarily contribute to global reference networks. By monitoring the ocean surface, satellite altimetry contributes to gravity field modeling, and high-resolution global gravity field recovery is expected from future satellite-to-satellite tracking and gravity gradiometry missions.

3.2.1 GPS

The Global Positioning System was designed and built, and is operated and maintained by the U.S. Department of Defense. The first GPS satellite was launched in 1978, and the system was fully operational in the mid-1990s. The GPS constellation consists of 24 satellites in six orbital planes with four satellites in each plane. The ascending nodes of the orbital planes are equally spaced by 60 degrees. The orbital planes are inclined 55 degrees. Each GPS satellite is in a nearly circular orbit with a semi-major axis of 26 578 km and a period of about twelve hours. The satellites continuously orient themselves to ensure that their solar panels stay pointed towards the Sun, and their antennas point toward the Earth. Each satellite carries four atomic clocks, is the size of a car and weighs about 1 000 kg. The long term frequency stability of the clocks reaches better than a few parts of 10^{-13} over a day. The atomic clocks aboard the satellite produce the fundamental L-band frequency, 10.23 MHz.

The GPS satellites are monitored by five base stations. The main base station is in Colorado Springs, Colorado and the other four are located on Ascension Island (Atlantic Ocean), Diego Garcia (Indian Ocean), Kwajalein and Hawaii (both Pacific Ocean). All stations are equipped with precise cesium clocks and receivers to determine the broadcast ephemerides and to model the satellite clocks. Transmitted to the satellites are ephemerides and clock adjustments. The satellites in turn use these updates in the signals that they send to GPS receivers.

Each GPS satellite transmits data on three frequencies: L1 (1 575.42 MHz), L2 (1 227.60 MHz) and L5 (1 176.45 MHz). The L1, L2 and L5 carrier frequencies are generated by multiplying the fundamental frequency by 154, 120 and 115, respectively. Pseudorandom noise (PRN) codes, along with satellite ephemerides, ionospheric model, and satellite clock corrections are superimposed onto the carrier frequencies L1, L2 and L5. The measured transmitting times of the signals that travel from the satellites to the receivers are used to compute the pseudoranges. The Course-Acquisition (C/A) code, sometimes called the Standard Positioning Service (SPS), is a pseudorandom noise code that is modulated onto the L1 carrier. The precision (P) code, sometimes called the Precise Positioning Service (PPS), is modulated onto the L1, L2 and L5 carriers allowing for the removal of the effects of the ionosphere.

The Global Positioning System (GPS) was conceived as a ranging system from known positions of satellites in space to unknown positions on land and sea, as well as in air and space. The orbits of the GPS satellites are available by broadcast or by the International Geodetic Service (IGS). IGS orbits are precise ephemerides after postprocessing or quasi-real time processing. All GPS receivers have an almanac programmed into their computer, which tells them where each satellite is at any given moment. The almanac is a data file that contains information of orbits and clock corrections of all satellites. It is transmitted by a GPS satellite to a GPS receiver, where it facilitates rapid satellite vehicle acquisition within GPS receivers. The GPS receivers detect, decode and process the signals received from the satellites to create the data of code, phase and Doppler observables. The data may be available in real time or saved for downloading. The receiver internal software is usually used to process the real time data with the single point positioning method and to output the information to the user. Because of the limitation of the receiver software, precise positioning and navigating are usually carried out by an external computer with more powerful software.

The basic contributions of the GPS are to tell the user where he is, how he moves, and what the timing is. Applications for GPS already have become almost limitless since the GPS technology moved into the civilian sector. Understanding GPS has become a necessity.

3.2.2 SLR

The laser has been adapted to measuring distances over the earth's surface and for computing ranges from earth stations to satellites and the lunar surface. The laser instrument is pointed to a target and then activated by a clock at the appropriate time. The laser beam is reflected at the target by special reflectors and the returning light is detected photoelectrically, and its time of flight measured to yield range data. The laser transmitter is mounted adjacent to some type of telescope or optical device used for receiving the reflected laser beam.

In satellite laser ranging, the interval between the outgoing and returning pulse from the satellite is measured very accurately and then transformed into a range measurement which is corrected for atmospheric refraction. Laser ranging is possible even when the satellite is in the earth's shadow and during daylight hours.

Simultaneous laser ranging to a near-earth satellite from two sites is used to determine the coordinates of one laser site relative to the fixed position of the other site and simultaneously the inter-site distance. NASA has used laser tracking since 1972 to measure the distance between points in North America. They have been testing the accuracy of laser tracking in measuring the crustal movement between points on opposite sides of the San Andreas fault and plan to make repeated measurements of baselines across the fault over a number of years. Simultaneous laser tracking has also been achieved between an east coast site and Bermuda enabling a determination of the Bermuda site's relative location (North American Datum) and the baseline between the two sites.

Laser ranging data has been incorporated into the development of world geodetic systems by the Smithsonian Astrophysical Observatory (SAO) and the Department of Defense (DoD). NASA has also included laser data in their development of gravitational models. Laser data is also being used for polar motion and earth rotation studies.

3.2.3 Satellite Altimetry

Satellite altimetry is based on a satellite-borne radar altimeter that transmits pulses in the vertical direction to the earth's surface. The ocean surface reflects the pulses perpendicularly, and the measurement of the travel time Δt furnishes the height of the satellite above the instantaneous sea surface (Fig. 3.1).

$$a = \frac{c}{2}\Delta t \qquad (3.1)$$

In spherical approximation, this observation can be expressed as

$$a = r_S - r_P - (N + \text{SST}) \qquad (3.2)$$

where r_S and r_P are the geocentric distances to the satellite and to the subsatellite point P on the ellipsoid; N is the geoid height and SST the height of the surface topography. Satellite tracking provides r_S and positioning gives r_P. Altimetry thus delivers information on the geoid and on sea surface topography.

Radar altimeters operate in the 14 GHz frequency range with short (a few ns) pulses and a high-pulse frequency (e.g., 100 pulses/s). The effects of beam divergence and finite pulse length result in measurements that refer to a "mean" sea surface within a circular "footprint" (few km diameter); Short-wavelength features of the ocean (waves) are thereby smoothed out. The observations must be corrected for instrumental effects (through calibration in the laboratory and in test areas), for the influence of the atmosphere, and for the oceanic tides (tide model). The "quasi-stationary" sea surface topography is obtained when considering the temporal variations of oceanographic and meteorological type, and in the water budget. With an accurate determination of the orbit, the position of this surface can be established in the geocentric coordinate system. On longer satellite missions, the sea surface can be covered several times with approximately uniform profiles.

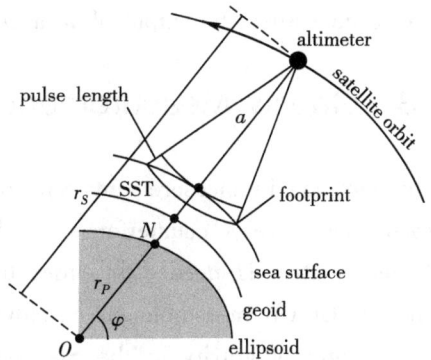

Fig. 3.1 Satellite altimetry

After the successful experiments on the Skylab mission, the first global survey with a radar altimeter was accomplished by the GEOS 3 satellite. The average over 2 seconds of measurements (footprint spacing of 15 km) is accompanied by an accuracy of ±0.5 m. This measuring accuracy has been increased in the Seasat 1 mission (1978) to ±0.1 m by changing to a shorter pulse length.

The Radar altimeters of the ERS-1 and the TOPEX/POSEIDON satellites deliver accuracies better than 10 cm.

3.2.4 Satellite-to-Satellite Tracking, Satellite Gravity Gradiometry

High-resolution gravity-field determination from space requires low-orbiting satellites and highly sensitive sensors. This is achieved by satellite-to-satellite tracking and satellite gravity gradiometry.

Satellite-to-satellite tracking (SST) employs microwave systems for measuring range rates between two satellites. High-low (one high and one low-flying satellite) and low-low (two low-flying satellites at the same altitude) configurations have been designed. The basic observables are the range rates (radial velocities) and changes of the range rates, which are due to gravitational and non-gravitational "disturbing" forces. The gravitational field parameters (harmonic coefficients) can be derived after proper compensation of the surface forces. In order to achieve a gravity field resolution of 100 km, the orbital altitude of the lower satellite must not exceed a few 100 km. The relative velocity between the satellites has to be determined with an accuracy of ±1 to 10μm/s, and precise tracking should be guaranteed by high-altitude satellite systems (GPS) and ground stations.

Satellite gravity gradiometry determines the components of the gravitational gradient tensor (second derivatives of the gravitational potential). On the earth's surface, gravity gradiometry has been employed since about 1900 with sensor pairs (accelerometers) sensitive to local changes of the gravity field in a certain direction. By different orientation of the sensors, different components of the gravity gradient can be determined. For space-borne applications, the attenuation of the gravity field with height (at a few 100 km height, the off-diagonal elements of the gradient tensor are a few 10^{-9} s^{-2} only) requires a high accuracy for the second derivatives (at the order of 10^{-11} to 10^{-13} s^{-2}), which can be achieved with conventional or superconducting electronics. High demands are posed on the attitude control of the sensor pairs and on the drift of the accelerometers. Surface forces cancel when the output of an accelerometer pair is differenced.

3.3 Gravity Measurements

"Absolute" gravity measurements refer directly to the standards of length and time, while "relative" measurements use a counterforce for the determination of gravity differences. A global gravity reference system is needed in order to refer local and regional gravity networks to a common standard. Gravity measurements on moving platforms are valuable for areas difficult to access.

The unit of gravity in the SI-system is ms^{-2}. The units $mGal = 10^{-5}\ ms^{-2}$ and $\mu Gal = 10^{-8}\ ms^{-2} = 10\ nms^{-2}$ are still in widespread use in geodesy and geophysics and derived from the unit Gal (after Galileo) of the former cgs-system.

At the earth's surface, g, the acceleration due to gravity
$$g \approx 980\ cm/s^2 \equiv 980\ Gal$$

Surveying and static gravity observations require accuracies of approximately 1 mGal (10^{-6} accuracy). To measure changes in gravity you'd like $1\mu Gal$ ($\sim 10^{-9}$) accuracies. For example, if you move radially outward by 3 mm, gravity decreases (because you are moving further from the center of the earth) by about $1\mu Gal$. So, $1\mu Gal$ accuracy corresponds to about 3 mm of vertical displacement. Most secular vertical crustal movement, when they occur, are on the order of up to a few mm per year. So a few years of gravity measurements at this accuracy level would allow those motions to be estimated.

There are two types of gravimeters:

Absolute meters Where g can be directly determined by measuring a length and/or a time.

Relative meters Where g depends on things like spring constants, which cannot be so readily determined. Relative instruments can only tell you the relative difference in g between two points or between two times.

The values of gravity required in geodesy and geophysics must refer to a global reference system. A gravity reference system is defined by values of gravity at a number of accurately surveyed gravity control points.

The **Potsdam Gravity System** served as the international reference system from 1909 to 1971. It was based on reversible pendulum measurements that were made at the Geodetic Institute at Potsdam (1898—1904). More recent absolute gravity determinations showed that the gravity value of Potsdam is 14 mGal too large. Therefore, between 1950 and 1970, a new global gravity system was constructed through international collaboration.

This **International Gravity Standardization Net 1971** (IGSN71) was introduced in 1971 as the new reference system, at the General Assembly of the IUGG in Moscow. The network contains 1 854 points ($\sim$ 500 primary stations) whose values of gravity were determined from ten new absolute and approximately 25 000 relative measurements of gravity (including $\sim$ 1 200 relative pendulum measurements) with an overall uncertainty less than ± 0.1 mGal. Particularly high accuracy is associated with points of the densely observed gravimeter calibration lines (Euro-African,

American, West Pacific) extending in the north-south direction (large differences in gravity). Regional gravity networks referred to other systems should be tied to the IGSN71. The transformation parameters (generally shift and scale) are derived from identical points, where data are available in both systems.

The IAG has proposed an International Absolute Gravity Basestation Network(IAGBN), with 36 globally distributed stations. Main purpose of this network is to monitor temporal gravity changes on a global scale, and to serve as a regional gravity control. Since 1987, the network is established with the help of transportable absolute gravimeters.

3.4 Astronomic Measurements

Classical geodetic astronomy is concerned with the determination of the astronomic latitude, longitude, and azimuth from ground-based optical direction measurements to fixed stars, which also requires time determination.

Geodetic astronomy is based on spherical astronomy. Its importance has decreased since the development of efficient satellite positioning and gravimetric methods and is now restricted to more local applications of gravity field (plumb line direction) and azimuth determinations. On the other hand, radio waves emitted from extragalactic sources are used extensively in order to derive base line vectors between fundamental terrestrial stations and to determine earth-rotation parameters: Very Long Baseline Interferometry.

The determination of astronomic latitude, longitude and azimuth is based on the relations given in geodetic astronomy, where the star positions are given by star catalogues. We ignore here the methods, such as Horrebow-Talcott method, developed in geodetic astronomy.

Observations of extragalactic radio sources such as quasars, can provide the geodetic information to determine the vector separations between the antennas of two widely separated radio telescopes. The components of the vector are its length and direction. To accomplish this, it is necessary to measure very accurately the difference in the time of arrival, recorded at the two antennas, of a particular wavefront from a given (point) source of radio radiation. The phenomena called interference, in Very Long Baseline lnterferometry (VLBI), is produced by electronically superimposing the recorded signals to produce a resultant disturbance or "interference" pattern. The theoretical expression for the relative phase delay shows it to be a function of the source direction, the antenna locations, the relative clock error between the two sites, the time of day, the model atmosphere employed, the earth's tidal parameters, the radio frequency at which the observation is made, etc. Proper account must also be taken of the earth's rotation. Two of the main limiting factors in the VLBI technique are clock stability and atmospheric variations. A major goal of VLBI is to reduce the uncertainty in intercontinental baselines to the centimeter level.

VLBI derived baselines have already contributed scale information to the development of the DoD World Geodetic System in 1972. Baselines accurate to the centimeter level would function as standards of comparison for future world systems. Other applications of VLBI include the determination of polar motion, variations in the earth's rotation, and the monitoring of motions of the major plates that compose the earth's crust.

Vocabulary

azimuth /ˈæzɪməθ/ n. 方位角
baseline /ˈbeɪslaɪn/ n. 基线
control network /kənˈtrəʊl ˈnetwɜːk/ 控制网
crustal movement /ˈkrʌstəlˈmuːvmənt/ 地壳运动
direction measurement /dɪˈrekʃən ˈmeʒəmənt/ 方向测量
ellipsoid /ɪˈlɪpsɔɪd/ n. 椭球
geoid height /ˈdʒiːɔɪd haɪt/ 大地水准面高
gravimeter /grəˈvɪmɪtə/ n. 重力仪
theodolite /θiˈɔdəlaɪt/ n. 经纬仪
total station /ˈtəʊtəl ˈsteɪʃən/ 全站仪
traverse /ˈtrævəs/ n. 导线
trigonometric leveling /ˌtrɪɡənəˈmetrɪk levəlɪŋ/ 三角高程测量
trilateration /traɪˌlætəˈreɪʃən/ n. 三边测量

4 Geodetic Networks

Geodetic and gravimetric networks consist of monumented control points that provide the reference frames for positioning and gravity-field determination at all scales. **Global networks** allow realization of the reference systems defined by international conventions. **Regional networks** form the fundamental basis for national or supranational (continental) geodetic and gravimetric surveys, which are the basis of geo-information systems and map series. **Local networks** are typically established for engineering and exploration projects and for geodynamic investigations.

In the sequel we concentrate on regional networks, which are increasingly integrated into global reference-frames established for positioning, navigation, and gravity. Horizontal and vertical control networks have been established separately, following the classical treatment of positioning and heighting; these networks still are the basis of national geodetic systems. Geodetic space methods allow establishment of three-dimensional networks within the geocentric reference system and are now superseding the classical control networks. Gravity networks serve the needs of geodesy and geophysics, with the reference provided either by a global network or by absolute gravimetry.

4.1 Horizontal Control Networks

National horizontal control networks were established from the 18 th century until the 1960's, with the networks' design, observation, and computation changing with the available techniques. Computations were carried out on a reference ellipsoid fitted to the survey area. Since the 1960's, spatial geodetic methods have allowed orientation of the classical networks with respect to the global reference system.

4.1.1 Design, Monumentation, Observations

Horizontal control networks are realized by trigonometric (triangulation) points, which in principle should be distributed evenly over the country. One distinguishes between different orders of trigonometric points, from first order or primary (station separation 30 to 60 km) to second order (about 10 km) to fourth or even fifth order (down to one to two km) stations. The maximum distance between first-order points was determined by the terrestrial measurement methods, which required intervisibility between the network stations. Consequently, first and some second-order stations were established on the top of hills and mountains; observation towers (wooden or steel constructions with heights of 30 m and more) were erected especially in flat areas. The stations have been permanently marked by underground and surface monuments (stone plates, stone or concrete pillars, bolts in hard bedrock). Eccentric marks have been set up in order to aid in the recovery and verification of the center mark.

Classical first and second-order horizontal control networks have been observed by the methods of triangulation, trilateration, and traversing.

In **triangulation**, all angles of the triangles formed by the trigonometric points are observed with a theodolite. The instrument is set up on the observation pillar or tower; at large distances the targets are made visible by light signals. Either directions (successive observation of all target points) or angles (separate measurement of the two directions comprising one angle) are observed in several sets (i. e., in both positions of the telescope), distributed over the horizontal circle of the theodolite. The scale of a triangulation network is obtained from the length of at least one triangulation side, either derived from a short base line through a base line extension net or measured directly by a distance meter. Astronomic observations provide the orientation of the network, whereas an astronomic azimuth is needed for the horizontal orientation according to Laplace's equation: Laplace station. In extended networks, base lines and Laplace stations often were established at distances of a few 100 km in order to control the error propagation through the network with respect to scale and orientation (effects of lateral refraction).

Trilateration employs electromagnetic distance meters in order to measure the lengths of all triangle sides of a network, including diagonals. At least one Laplace azimuth is needed for the orientation of the net. Electromagnetic distance measurements put less demands on the stability of observation towers as compared to angular measurements, and the use of laser light and microwaves makes the method more independent from weather conditions.

Traverses combine distance and angular measurements, where the traverse stations are arranged along a profile. Again, a Laplace azimuth is required in order to orientate the traverse. Traversing represents a very effective and flexible method for establishing horizontal control networks, with no more need to establish stations on hilltops. It has been employed primarily for the densification of higher order networks, with the coordinates of the existing control points providing orientation.

Horizontal control networks can also be formed by combining the methods of triangulation, trilateration, and traversing.

4.1.2 Computation and Orientation

Classical first and some second-order horizontal control networks have been calculated on a reference ellipsoid within the system of ellipsoidal coordinates. Lower order networks are primarily calculated in planar Cartesian coordinates, after conformal mapping of the ellipsoid onto the plane.

The observed horizontal angles, directions, and spatial distances were reduced to the ellipsoid first; the gravity-field-related reductions (deflections of the vertical, geoid height) were not considered during earlier surveys. The network adjustment was carried out either by the method of conditions or by variation of the coordinates, with redundancy resulting from triangle misclosures, diagonals in trilateration quadrilaterals, and additional base lines and Laplace azimuths. For the method of conditions, the geometry of the network was adjusted first. The coordinates transfer from an **geodetic origin** was then carried out using the solutions of the direct problem on the ellipsoid;

while for the method of variation of the coordinates, ellipsoidal "observation equations" were derived from the solution of the inverse problem. Among the deficiencies of this classical "**development method**" are the neglecting of the deflections of the vertical, the inadequate reduction of distances on the ellipsoid, and especially the step by step calculation of larger networks, with junction constraints when connecting a new network section to an existing one. This led to network distortions of different type, with regionally varying errors in scale ($\pm 10^{-5}$) and orientation ($\pm$ a few arcsec). Relative coordinate errors with respect to the geodetic origin increased from a few decimeters over about 100 km to about one meter over several 100 km and reached 10 meters and more at the edges of extensively extended networks.

4.2 Vertical Control Networks

Vertical control networks have been established separately from horizontal control nets. This is due to the demand that heights have to refer to the gravity field rather than the ellipsoidal system used for horizontal positioning.

Vertical control networks are determined by geometric leveling and occasionally by hydrostatic leveling, the control points being designated as bench marks. According to the leveling procedure and the accuracy achieved, national geodetic surveys distinguish between different orders of leveling. First-order leveling is carried out in closed loops (loop circumferences of some 100 km) following the rules for precise leveling. The loops are composed of leveling lines connecting the nodal points of the network. The lines, in turn, are formed by leveling runs that connect neighboring benchmarks. The first-order leveling network is densified by second to fourth-order leveling.

Leveling lines generally follow main roads, railway lines, and waterways. The benchmarks consist of bolts in building, bedrock, or on concrete posts. Long pipes are set up in alluvial regions. Underground monuments are established in geologically stable areas in order to secure the network stability with respect to variations with time. An accuracy of about 1 mm per km is achieved for first-order networks. They should be reobserved at time intervals of some 10 years, as regional and local height changes can reach 1 mm/year and more, especially in areas which experience crustal movement.

Prior to the adjustment of a leveling network, the observed raw height differences have to be transformed either to geopotential differences or to differences of **normal or orthometric heights** by considering surface gravity. The adjustment then utilizes the loop misclosure condition of zero and is carried out either by the method of condition equations or, preferably, by the method of parameter variation.

The **vertical datum** (zero height surface) is defined by **mean sea level** (MSL) as derived from tide gauge records. National height systems may differ by some dm to 1 m, and more, between each other. They may also differ from the geoid as a global reference surface, which is due to the effect of sea surface topography. Network distortions arise if the vertical datum is constrained to MSL of more than one tide gauge.

4.3　Three-dimensional Networks

Geodetic space methods deliver three-dimensional coordinates with cm-accuracy in the global geocentric system on global, regional, and local scale. Global networks realize the terrestrial reference system and provide reference stations for the establishment of continent-wide networks, wherein GPS plays a fundamental role. Further network densification is carried out nearly exclusively by GPS, with subsequent integration of existing classical control nets.

The ITRF represents the global basis for three-dimensional positioning. It is defined by the geocentric Cartesian coordinates and the horizontal velocities of a global set of space geodetic observing sites, given for a certain epoch, with an accuracy of about ±1 cm and ±1 to 3 mm/year respectively. The ITRF results are based on observations obtained by global networks employing different space techniques, which are combined within the frame of the IERS.

National geodetic control-networks have been established by GPS since the 1990's and now supersede the classical horizontal (and possibly vertical) control networks. Although the strategies for establishing and maintaining these thee-dimensional reference networks are still under discussion and differ from country to country, the following directions clearly can be identified:

(1) Establishment of a large-scale, three-dimensional network with station distances between 10 and 50 km.

(2) Installation of permanent GPS stations with distances of 50 to 100 km.

(3) Transformation of the existing horizontal control network, eventually using additional GPS control.

4.4　Gravity Networks

Gravity networks provide the frame for gravimetric surveys on global, regional, or local scales. They consist of gravity stations where gravity has been determined by absolute or relative methods. On a global scale, the gravity standard is realized by the IGSN71, but absolute gravimeters now allow independent establishment of the gravity standard. A global, absolute-gravity base-station network has been established primarily for the investigation of long-term gravity variations with time.

National gravimetric surveys are based on a primary or base network, which in most cases is densified by lower order nets. The gravity base-network stations should be evenly distributed over the area; station distances of a few 100 km are typical for larger countries. The station sites should be stable with respect to geological, hydrological, and microseismic conditions, and they should be located at permanent locations (observatories etc.). Eccentric stations serve for securing the center station and for controlling local height and mass changes. The stations of the subsequent densification networks may be collocated with horizontal and vertical control points. Horizontal position and height of the network stations should be determined with m and mm to cm-accuracy respectively.

Vocabulary

development method /dɪˈveləpmənt ˈmeθəd/ 展平法
horizontal control network /ˌhɔrɪˈzɔntəl kənˈtrəul ˈnetwəːk/ 平面控制网
geodetic origin /ˌdʒiːədetik ˈɔridʒin/ 大地原点
Laplace azimuth /lɑːˈplɑːs ˈæziməθ/ 拉普拉斯方位角
mean sea level (MSL) /miːn siː ˈlevl/ 平均海水面
orthometric height /ɔːˈθɔmitrik hait/ 正高
reference ellipsoid /ˈrefrəns iˈlipsɔid/ 参考椭球
vertical datum /ˈvəːtikəl ˈdeitəm/ 高程基准

5 The gravity field of the Earth

The external gravity field plays a fundamental role in geodesy. That is because the figure of the earth has evolved under the influence of gravity, and most geodetic observations refer to gravity. Modeling of the observations thus requires knowledge of the gravity field. In addition, the analysis of the gravity field yields information on the structure of the earth's interior; in this way geodesy contributes to geophysics.

5.1 Fundamentals of the Gravity Field

A body rotating with the earth experiences the gravitational force of the masses of the earth as well as the centrifugal force due to the earth's rotation. The resultant is the force of gravity. In the case of artificial satellites, it is noted that a satellite does not rotate with the earth; hence, only gravitation acts on the satellite.

5.1.1 Gravitation, Gravitational Potential

According to Newton's law of gravitation, two point masses m_1 and m_2 attract each other with the gravitational force (attractive force)

$$\boldsymbol{F} = -G \frac{m_1 m_2}{r^2} \frac{\boldsymbol{r}}{r} \tag{5.1}$$

where G is the gravitational constant, and r is the distance between the masses. The vectors $\boldsymbol{F}$ and $\boldsymbol{r}$ point in opposing directions.

The representation of gravitational acceleration, the gravity field, and related computations are simplified if the scalar quantity "potential" is used instead of the vector quantity "acceleration".

For a point mass m, we have

$$V = \frac{Gm}{r} \tag{5.2}$$

For the earth, we obtain

$$V = G \iiint_{\text{earth}} \frac{dm}{r} = G \iiint_{\text{earth}} \frac{\rho}{r} dv \tag{5.3}$$

The **potential** at P indicates the work that must be done by gravitation in order to move the unit mass from infinity ($V = 0$) to P. The unit of potential is $m^2 s^{-2}$.

5.1.2 Centrifugal Acceleration, Centrifugal Potential

The centrifugal force acts on an object of mass on the earth's surface. It arises as a result of the rotation of the earth about its axis. We assume here a rotation of constant angular velocity ω about the rotational axis, with the axis assumed fixed with the earth. The centrifugal acceleration

$$z = (\boldsymbol{\omega} \times \boldsymbol{r}) \times \boldsymbol{\omega} = \omega^2 \boldsymbol{p} \tag{5.4}$$

acting on a unit mass is directed outward and is perpendicular to the spin axis (Fig. 5.1).

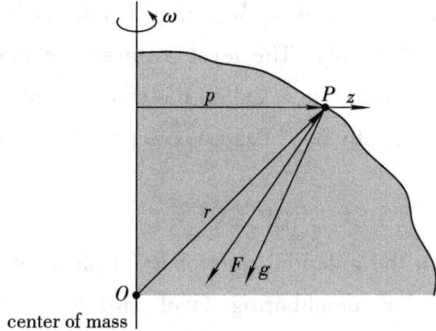

Fig. 5.1 Gravitation, centrifugal acceleration, and gravity

With

$$z = \text{grad } Z \tag{5.5}$$

We introduce the centrifugal potential

$$Z = \frac{\omega^2}{2} p^2 \tag{5.6}$$

5.1.3 Gravity Acceleration, Gravity Potential

The gravity acceleration, or gravity g, is the resultant of gravitation F and centrifugal acceleration z:

$$g = F + z \tag{5.7}$$

The direction of g is referred to as the direction of the plumb line (vertical); the magnitude g is called gravity intensity (often just gravity). With (5.3) and (5.6), the gravity potential of the earth becomes.

$$W = V + Z = G \iiint_{\text{earth}} \frac{\rho}{r} dv + \frac{\omega^2}{2} p^2 \tag{5.8}$$

It is related to the gravity acceleration by

$$g = \text{grad} W \tag{5.9}$$

5.2 Level Surfaces and Plumb Lines

The surfaces of constant gravity potential

$$W = W(\boldsymbol{r}) = \text{const.} \tag{5.10}$$

are designated as **equipotential** or **level surfaces** (also geopotential surfaces) of gravity. As a result of an infinitesimal displacement $d\boldsymbol{s}$, the potential difference of differentially separated level surfaces is given by

$$dW = \boldsymbol{g} \cdot d\boldsymbol{s} = g ds \cos(\boldsymbol{g}, d\boldsymbol{s}) \tag{5.11}$$

This means that the derivative of the gravity potential in a certain direction is equal to the component of the gravity along this direction. Since only the projection of ds along the plumb line enters into (5.11), dW is independent of the path. Hence, no work is necessary for a displacement along a level surface W = const.; the level surfaces are equilibrium surfaces.

If ds is taken along the level surface $W = W_P$, then it follows from dW = 0 that $\cos(\boldsymbol{g}, d\boldsymbol{s}) = \cos 90° = 0$: gravity is normal to $W = W_P$. The level surfaces are intersected at right angles by the plumb lines. The tangent to the plumb line is called the direction of the plumb line. If ds is directed along the outer surface normal n, then, because $\cos(\boldsymbol{g}, \boldsymbol{n}) = \cos 180° = -1$, the following **important relationship** exists:

$$dW = -g dn \tag{5.12}$$

It provides the link between the potential difference (a physical quantity) and the difference in height (a geometric quantity) of neighboring level surfaces. According to this relation, a combination of gravity measurements and differential height determinations, as provided by geometric leveling, delivers gravity potential differences.

5.3 Spherical Harmonic Expansion of the Gravitational Potential

Because the density function $\rho = \rho(\boldsymbol{r}')$ of the earth is not well known, the gravitational potential $V = V(\boldsymbol{r})$ cannot be computed by Newton's law of gravitation. However, a convergent series expansion of V is possible in the exterior space as a special solution of Laplace's differential equation. This solution corresponds to a spectral decomposition of the gravitational field. The coefficients of the series expansion provide the amplitudes of the respective spectral parts. Any observable functional of V can be evaluated for the determination of these coefficients, thus allowing a global analytical representation of the gravitational field.

The gravitational potential expanded in spherical harmonics is written as

$$V = \frac{GM}{r}\left[1 + \sum_{l=1}^{\infty}\sum_{m=0}^{l}\left(\frac{a}{r}\right)^l (C_{lm}\cos m\lambda + S_{lm}\sin m\lambda) P_{lm}(\cos\vartheta)\right] \tag{5.13}$$

where (r, ϑ, λ) are the spherical coordinates of point P, a is the semimajor axis of the earth ellipsoid, C_{lm}, S_{lm} are spherical harmonic coefficients, and $P_{lm}(.)$ the associated Legendre functions of the first kind. The harmonic coefficients are given by

$$\left.\begin{array}{l} C_{l0} = C_l = \dfrac{1}{M}\iiint\limits_{\text{earth}}\left(\dfrac{r'}{a}\right)^l P_l(\cos\vartheta')\,dm \quad\text{and} \\[2ex] \left\{\begin{array}{l}C_{lm}\\S_{lm}\end{array}\right\} = \dfrac{2}{M}\times\dfrac{(l-m)!}{(l+m)!}\iiint\limits_{\text{earth}}\left(\dfrac{r'}{a}\right)^l P_{lm}(\cos\vartheta')\left\{\begin{array}{l}\cos m\lambda'\\\sin m\lambda'\end{array}\right\}dm \quad\text{for } m\neq 0 \end{array}\right\} \tag{5.14}$$

where $(r', \vartheta', \lambda')$ are the coordinates of dm.

5.4 The Geoid

The geoid is of fundamental importance for geodesy, oceanography, and physics of the solid earth. In geodesy and oceanography, the geoid serves as a height reference surface for describing continental and sea surface topography. Geophysics exploits the geoid as a gravity field representation revealing the distribution of deeper located masses.

5.4.1 Definition

The **equipotential surface** in the gravity field of the earth which coincides with the undisturbed mean sea level extended continuously through the continents. The direction of gravity is perpendicular to the geoid at every point. The geoid is the surface of reference for astronomical observations and for geodetic leveling.

$$* * * * * * * * * *$$

We consider the waters of the ocean as freely moving homogeneous matter, which is subject only to the force of gravity of the earth. Upon attaining a state of equilibrium, the surface of such idealized oceans assumes a level surface of the gravity field; We may regard it as being extended under the continents (e. g. by a system of communicating tubes). This level surface is termed the **geoid**. Its equation is given by

$$W = W(r) = W_0 \qquad (5.15)$$

We see that the geoid is a closed and continuous level surface which extends partially inside the solid body of the earth. The curvature of the geoid displays discontinuities at abrupt density variations. Consequently, the geoid is not an analytic surface, and it is thereby eliminated as a reference surface for position determinations. However, it is well suited as a reference surface for potential or height differences, which are supplied by spirit (geometric) leveling in combination with gravity measurements.

5.4.2 Mean Sea Level

The average height of the surface of the sea for all stages of the tide, usually determined by averaging height readings observed hourly over a minimum period of 19 years. Also called sea level datum.

$$* * * * * * * * * *$$

In order to establish the geoid as a reference surface for heights, the ocean's water level is registered and averaged over longer intervals using tide gauges (mareographs). The **mean sea level** (MSL) thus obtained represents an approximation to the geoid.

Since the tide gauge stations usually do not have an undisturbed link to the waters of the oceans, the recordings are frequently falsified by systematic influences. The variations of the sea level with time, as long as they are periodic, are largely eliminated by averaging the water level registrations.

Satellite altimetry furnishes data on the open seas which refers to the instantaneous water surface. The height of the ocean surface above the geoid represents the sea surface topography. Here, one distinguishes between the instantaneous sea surface topography and the quasistationary sea surface topography which results after accounting for the time dependent variations.

These variations include ocean tides which can deviate considerably from the theoretical values due to unequal water depths and because the continents impede the movement of water. The tidal bulge on the open sea is less than 1 m; however, it can amount to several meters in coastal areas. Fluctuations which usually have yearly periods and attain values up to 1 m include those of a meteorological nature (atmospheric pressure, winds), those of an oceanographic nature (ocean currents, differences in water density as a function of temperature, salinity, and pressure), and those due to the water budget (changing water influx resulting from meltwater, monson rains, etc.). At present, a secular variation of about 1 mm per year is caused by the postglacial increase in sea level.

Although the internal accuracy of the average annual values of the water level observations amounts to ±1 cm, occasional deviations of ±10 cm and higher may occur among the yearly averages (meteorological effects).

Even after comprehensively including all periodic, nonperiodic, and secular variations, which is possible only in exceptional cases, the mean sea level yet does not form a level surface of the earth's gravity field. Over larger areas, the deviations can amount to 1 m and more.

5.5 Height Systems

The geoid is used in geodesy, cartography, and oceanography as a reference surface for heights and depths (continental and ocean bottom topography, as well as sea surface topography). A point P can be attributed to a specified level surface by its gravity potential W (Fig. 5.2). With respect to the geoid potential W_0, the "height" of P is given by the negative potential difference to the geoid, which is called the **geopotential number** C. We get from (5.12)

$$C = W_0 - W_P = -\int_{P_0}^{P} \mathrm{d}W = \int g \mathrm{d}n \tag{5.16}$$

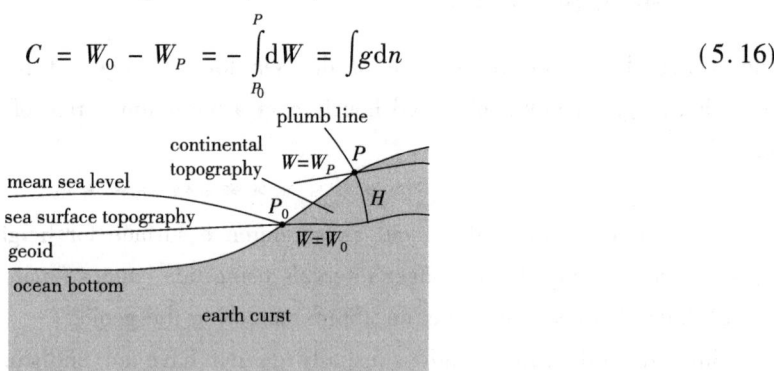

Fig. 5.2 Geoid, mean sea level, continental and sea surface topography

The integral is independent of the path; hence, P_0 is an arbitrary point on the geoid. C can be determined from geometric leveling and gravity measurements along any path between P_0 and P.

The geopotential number is an ideal measure for describing the behavior of masses (e.g., water masses) in the gravity field. It could be used as a "height" in several applications, as in hydraulic engineering and oceanography. A more general use is limited by the potential unit $m^2 s^{-2}$, which is in contradiction to the obvious demand for a metric height system that employs the "meter" unit.

In order to achieve a certain agreement with the numerical value of the height in meters, the geopotential unit (gpu) $10\text{ m}^2\text{s}^{-2}$, or kgalm, is also used for the geopotential number. With $g \approx 9.80\text{ ms}^{-2}$, the values of C are about 2% smaller than the corresponding height values.

The **dynamic height** H^{dyn} is obtained by dividing the geopotential number through a constant gravity value. Usually the normal gravity γ_0^{45} calculated for the surface of the level ellipsoid at 45° latitude is used:

$$H^{\text{dyn}} = \frac{C}{\gamma_0^{45}} \tag{5.17}$$

The surfaces H^{dyn} = const. remain equilibrium surfaces. Hence, points located on the same level surface have the same dynamic height. Unfortunately, a geometric interpretation of the dynamic heights is not possible, and larger corrections are necessary in order to convert leveling results into dynamic height differences. Because of this, dynamic heights have not been widely used in geodesy but are used in oceanography.

National or continental height systems, and terrain-data based on them (topographic maps, digital terrain models), use either orthometric or normal heights.

The **orthometric height** H is defined as the linear distance between the surface point and the geoid, reckoned along the curved plumb line. This definition corresponds to the common understanding of "heights above sea level". Expanding the right-hand side of (5.16) in H and integrating along the plumb line from $P_0(H=0)$ to $P(H)$ we obtain

$$H = \frac{C}{\bar{g}}, \quad \bar{g} = \frac{1}{H}\int_0^H g\,\mathrm{d}H \tag{5.18}$$

$\bar{g}$ is the mean gravity along the plumb line; gravity values inside the earth are required for its calculation. This is performed by introducing a model of the density distribution of the topographic masses. As this distribution is known only imperfectly, the accuracy of computed orthometric heights depends on the accuracy of the density model. In addition, points of equal orthometric height deviate slightly from a level surface, which is due to the non-parallelism of the level surfaces. These drawbacks are compensated by the fact that orthometric heights represent the geometry of the topographic masses. Geometric leveling only needs small corrections for the transformation into orthometric height differences.

In order to avoid any hypothesis on the distribution of the topographic masses, **normal heights** H^N have been introduced and are used in a number of countries. The mean gravity $\bar{g}$ in (5.18) is replaced by the mean normal gravity $\bar{\gamma}$ along the normal plumb line, which is only slightly curved:

$$H^N = \frac{C}{\bar{\gamma}}, \quad \bar{\gamma} = \frac{1}{H^N}\int_0^{H^N} \gamma\,\mathrm{d}H^N \tag{5.19}$$

$\bar{\gamma}$ can be calculated in the normal gravity field of an ellipsoidal earth model. The reference surface for the normal heights is the **quasi-geoid**, which is close to the geoid but not a level surface. It deviates from the geoid on the mm to cm-order at low elevations and may reach one-meter deviation in the high mountains. On the oceans, geoid and quasi-geoid practically coincide.

The zero height surfaces (vertical datum) of national height systems generally is defined by the mean sea level derived from tide gauge records over a certain time interval. These reference surfaces only approximate the geoid, due to the sea surface topography and local anomalies, which deviations up to one meter.

Vocabulary

centrifugal /sen'trifjəgəl, -'trifə-/ *adj.* 离心的
dynamic height /dai'næmik hait/ 力高
earth ellipsoid /ə:θ i'lipsɔid/ 地球椭球
equipotential /ˌi:kwipəu'tenʃəl/ *adj.* 等位的
geopotential number /ˌdʒi:əupə'tenʃəl 'nʌmbə/ 地球位数
Legendre functions /lə'ʒɑ:ŋdə 'fʌŋkʃənz/ 勒让德函数
normal gravity /'nɔ:məl 'græviti/ 正常重力
normal height /'nɔ:məl hait/ 正常高
quasi-geoid /ˌkwə:zi(:) 'dʒi:ɔid/ *n.* 似大地水准面
spherical harmonic expansion /'sfiərikəl hɑ:'mɔnik iks'pænʃən/ 球谐展开

6 Reference Ellipsoid and Geodetic Coordinate System

A geodetic earth model is used as a reference for the actual surface of the earth. It should provide a good fit to the geoid and thus allow the linearization of non-linear geodetic problems. On the other hand, the mathematical formation of the model should be simple and possibly permit calculations by closed formulas. The model should serve as a standard for applications not only in geodesy and cartography but also in astronomy and geophysics; it should satisfy the demands and needs of these disciplines too.

6.1 The Rotational Ellipsoid

The rotational ellipsoid was introduced as a geometrical figure of the earth in the 18th century. The geometry of the ellipsoid can be described in a simple manner, together with ellipsoidal surface coordinates and curvature.

6.1.1 Geometric Parameters

The rotational ellipsoid is created by rotating the meridian ellipse about its minor axis. The shape of the ellipsoid is thereby described by two geometric parameters; the **semimajor axis** a and the **semiminor axis** b. Generally, b is replaced by one of a number of smaller quantities which is more suitable for series expansion: the (geometrical) **flattening** f, the **linear eccentricity** ε, the **first and second eccentricities** e and e', respectively:

$$f = \frac{a-b}{a}, \quad \varepsilon = \sqrt{a^2 - b^2}, \quad e = \frac{\varepsilon}{a}, \quad e' = \frac{\varepsilon}{b} \tag{6.1}$$

The following relations hold among these quantities:

$$\frac{b}{a} = 1 - f = \sqrt{1 - e^2} = \frac{1}{\sqrt{1 + e'^2}} \tag{6.2}$$

6.1.2 Geodetic and Cartesian Coordiantes

An alternative and often more convenient method of defining position is Cartesian coordinates.

The system has its origin at the center of the earth with the X and Y axes in the plane of the equator. The Z axis coincides with the mean rotation axis of the Earth (for example, the Z axis has the direction of the CIO), and the X axis passes through the zero meridian, which is the mean Greenwich meridian (the zero meridian adopted by BIH). The three axes are mutually orthogonal and form a right-handed system.

The geodetic coordinates B (geodetic latitude), L (geodetic longitude), and h (height above the reference ellipsoid) are then related to the Cartesian coordinates XYZ by the well-known equations:

$$X = (N + h)\cos B\cos L$$
$$Y = (N + h)\cos B\sin L \quad (6.3)$$
$$Z = [N(1 - e^2) + h]\sin B$$

where N is the **radius of curvature in the prime vertical** (east-west direction):

$$N = a/\sqrt{1 - e^2\sin^2 B} \quad (6.4)$$

Cartesian coordinates are usually used as the computational system of coordinates in Doppler positioning.

The inverse transformation can be solved by iterations or in closed form. Both approaches employ the distance P from the minor axis which, for any point, equals

$$P = \sqrt{X^2 + Y^2} \quad (6.5)$$

or, from (6.3),

$$P = (N + h)\cos B \quad (6.6)$$

From (6.3) and (6.6) we have

$$\frac{Z}{P} = \tan B\left(1 - \frac{e^2 N}{N + h}\right) \quad (6.7)$$

This equation is the point of departure for both approaches. The iterations are usually initiated by solving first for B from the above equation. Putting $h = 0$, we get

$$B^{(0)} = \arctan\left(\frac{Z}{P}\frac{1}{1 - e^2}\right) \quad (6.8)$$

The Kth iteration then consists of evaluating successively $N^{(K)} = N(B^{(K-1)})$ from $N = a/\sqrt{1 - e^2\sin^2 B}$; $h^{(K)} = h(B^{(K-1)}, N^{(K)})$ from (6.6); and $B^{(K)} = B(N^{(K)}, h^{(K)})$ from (6.7). The iterations are repeated until the following inequalities are satisfied:

$$|h^{(K)} - h^{(K-1)}| < a\varepsilon \text{ and } |B^{(K)} - B^{(K-1)}| < \varepsilon \quad (6.9)$$

for some a priorly chosen value of ε. Once B and h are found, L is evaluated from either of the first two equations (6.3) or

$$L = 2\arctan\frac{Y}{X + \sqrt{X^2 + Y^2}} \quad (6.10)$$

The closed form solution uses (6.6) and (6.3) to obtain

$$P\tan B - Z = e^2 N\sin B \quad (6.11)$$

In this equation, the only unknown is B, N being a function of B as well. Substituting from N, (6.11) changes to

$$P\tan B - Z = \frac{ae^2\sin B}{\sqrt{1 - e^2\sin^2 B}} \quad (6.12)$$

Dividing the numerator and denominator of the right-hand side by $\cos B$ and squaring the whole equation yields

$$P^2\tan^4 B - 2PZ\tan^3 B + \left(Z^2 + \frac{P^2 - a^2 e^4}{1 - e^2}\right)\tan^2 B - \frac{2PZ}{1 - e^2}\tan B + \frac{Z^2}{1 - e^2} = 0 \quad (6.13)$$

This is a quartic (biquadratic) equation in tan$B$ in which the values of all the coefficients are known. Standard procedures for solving quartic equations exist. Once a solution for tan B is obtained, N and h are computed from $N = a/\sqrt{1 - e^2\sin^2 B}$ and (6.6) respectively. Longitude L follows directly from (6.3) or (6.10) thus completing the inverse transformation. Paul has shown that the closed form approach is about 25% faster than the iterative.

6.1.3 Projection onto the Ellipsoid

Let us establish the position of a point P by means of the nature coordinates φ, λ, H. Then we may project it onto the geoid along the (slightly curved) plumb line. The orthometric height is the distance between P and its projection P_o onto the geoid, measured along the plumb line (Fig. 6.1). Although this mode of projection is entirely natural, the geoid is not suited for performing computations on it directly; the point P_o is therefore projected onto the reference ellipsoid by means of the straight ellipsoidal normal, thus getting a point Q_o on the ellipsoid. In this way, the ground point P and the corresponding point Q_o on the ellipsoid are connected by a double projection, that is, by two projections which are performed one after the other and which are quite analogous, the orthometric height $H = PP_0$ corresponding to the geoidal undulation $N = P_0Q_0$. This double projection is called **Pizzetti's projection**.

It is simpler to project the point P from the physical surface of the earth directly onto the ellipsoid through the straight ellipsoidal normal, thus obtaining a point Q. The distance $PQ = h$ is the geometric height above the ellipsoid. The ground point P is then determined by h and the geographical coordinates B, L of Q on the ellipsoid, so that the so-called geodetic coordinates B, L, h take the place of the natural coordinates φ, λ, H. This projection is called **Helmert's projection**.

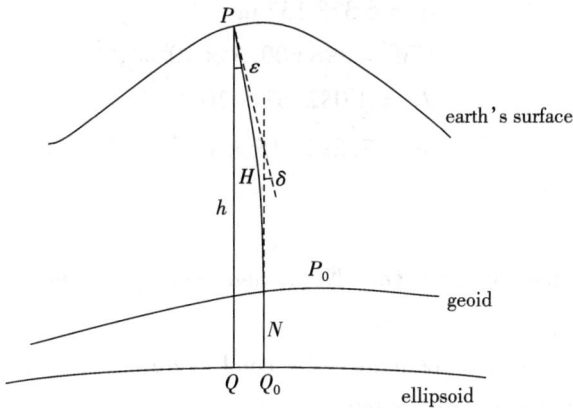

Fig. 6.1 The projection of Helmert and of Pizzetti

The practical difference between Pizzetti's and Helmert's projection is small. The ellipsoidal height h is equal to $H + N$ within a fraction of a millimeter. The geodetic coordinates B and L, with respect to the two projections, are related by the equations

$$B_{\text{Helmert}} = B_{\text{Pizzetti}} + \frac{H}{R}\xi$$
$$L_{\text{Helmert}} = L_{\text{Pizzetti}} + \frac{H}{R}\eta \sec B \tag{6.14}$$

which can be read from Fig. 6.1, since $QQ_0 = H\varepsilon$; $R = 6\,371$ km is the mean radius of the earth. Even if $\varepsilon = 1$ min of arc and $H = 1\,000$ m, the distances QQ_0 is only about 30 cm and the geodetic coordinates differ by less than $0.01''$, which is below the accuracy of astronomical observations. For most purposes we may therefore neglect the difference between the two projections.

Pizzetti's projection is better adapted to the geoid, because there is an exact correspondence between a geoidal point P_0 and an ellipsoidal point Q_0. Helmert's projection has practical advantages, notably the straightforward conversion of the ellipsoidal coordinates B, L, h into rectangular coordinates X, Y, Z; it is also simpler in other respects. For this reason we shall henceforth use mainly Helmert's projection, but practically the results hold for both projections.

However, because of the curvature of the plumb line, we have to distinguish carefully whether the astronomical coordinates refer to the ground point P or the geoidal point P_0. Even if the angle δ in Fig. 6.1 is only 1 second of arc, a change of $1''$ in the geographical latitude means a linear displacement of P by $R\delta = 30$ m. This must be taken into account if we combine astronomical coordinates φ and λ, measured at the ground point P and the gravimetric deflections of the vertical ξ and η, computed by Vening Meinesz' formula for the geoidal point P_0.

6.1.4 Numerical Values of Ellipsoid Parameters

The Geodetic Reference System 1980 has the following four defining parameters:

$$a = 6\,378\,137 \text{ m}$$
$$GM = 398\,600.5 \times 10^9 \text{m}^3\text{s}^{-2}$$
$$J_2 = 1\,082.63 \times 10^{-6}$$
$$\omega = 7.292\,115 \times 10^{-5} \text{rads}^{-1}$$

where

a = semimajor axis,

GM = geocentric gravitational constant (Newtonian constant G times mass M of earth including atmosphere),

J_2 = zonal spherical-harmonic coefficient of second degree,

ω = angular velocity of the earth's rotation.

From these defining constants, other parameters can be unambiguously derived, for instance, the flattening

$$f = 1/298.257\,222\,101$$

and equatorial gravity

$$\gamma_e = 9.780\,326\,771\,5 \text{ms}^{-2}$$

The Department of Defense World Geodetic System 1984 has

$$a = 6\ 378\ 137 \text{ m}$$
$$f = 1/298.257\ 223\ 563$$
$$GM = 3.986\ 004\ 418 \times 10^{14} \text{m}^3\text{s}^{-2}$$
$$\omega = 7.292\ 115 \times 10^{-5} \text{rad} \cdot \text{s}^{-1}$$

IAG plans to revise fundamental geodetic constants, including earth ellipsoid parameters, every 4 years, adopting a list of such values, currently considered representative, at each General Assembly. On the other hand, a geodetic reference system by no means needs to incorporate the currently best values. In fact, if all is correctly done, the final results (spatial position of triangulation points, the geoid, etc.) will be independent of the particular choice of the ellipsoidal reference system (as long as it is within reasonable limits), which thus plays only an intermediate role without having any effect on the final result. So the desire to have a reference ellipsoid that fits the geoid as closely as possible is motivated primarily by esthetical reasons. Practically much more important is long-term stability; an enormous amount of data is based on an adopted reference system, and such a reference should be changed as infrequently as possible.

6.1.5 Triaxial ellipsoid

An ellipsoid which has three mutually perpendicular unequal axes. It requires three parameters for definition. Geodetic calculations using it are unnecessarily complicated, and hence its use in geodesy should be avoided.

$$* * * * * * * * * * *$$

A mathematically viable surface, closest to the geoid, is a triaxial ellipsoid. Many researchers have estimated the parameters of a triaxial ellipsoid that would best approximate the geoid. Such a triaxial ellipsoid has three mutually perpendicular axes positioned in the earth as follow: the minor, coinciding with the earth's principal (polar) axis of inertia; the major, and the medium, both lying in the equatorial plane. Thus, the triaxial ellipsoid is defined by the lengths of the major axis ($2a$), the minor axis ($2b$), the medium axis ($2c$), and the orientation of the major axis in the equatorial plane.

Usually, the following are taken as the four defining parameters:
(a) Half length of the major axis a.
(b) Polar flattening f, given by $f = (a - b)/a$.
(c) Equatorial flattening f_e, given by $f_e = (a - c)/a$.
(d) Geographical longitude λ_a of the major axis.

Since the deviations of the biaxial rotational ellipsoid from the geoid (< 100 m) generally attain this order of magnitude, the triaxial ellipsoid does not present a considerably better fit to geoid and the gravity field. In contrast, the geodetic computations are encumbered by the intricate geometry. Lastly, the triaxial ellipsoid also is not suitable as a physical normal figure. The triaxial ellipsoid thus is not appropriate as a reference body, with the exception of special purposes.

6.2 Curvature

6.2.1 Curvature of the Rotational Ellipsoid

We introduce a spatial X, Y, Z Cartesian coordinate system (Fig. 6.2).

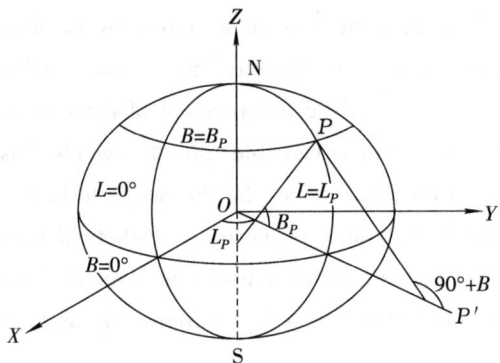

Fig. 6.2 Geodetic coordinates latitude and longitude

The origin of the system is situated at the center O of the figure, the Z-axis coincides with the minor axis of the ellipsoid. The equation of the surface of the ellipsoid is then given by

$$\frac{X^2 + Y^2}{a^2} + \frac{Z^2}{b^2} - 1 = 0 \tag{6.15}$$

We introduce the radius of the circle of latitude

$$P = \sqrt{X^2 + Y^2} \tag{6.16}$$

as a new variable. Substituting this into (6.15) and differentiating yields the slope of the ellipsoidal tangent at P

$$\frac{dZ}{dP} = -\left(\frac{b}{a}\right)^2 \frac{P}{Z} = -\cot B \tag{6.17}$$

From (6.15) and (6.17) the parametric representation of the meridian ellipse follows:

$$P = \frac{a^2 \cos B}{\sqrt{a^2 \cos^2 B + b^2 \sin^2 B}} \tag{6.18}$$

$$Z = \frac{b^2 \sin B}{\sqrt{a^2 \cos^2 B + b^2 \sin^2 B}}$$

The meridians and parallels are the lines of curvature of the rotational ellipsoid. The principal radii of curvature are therefore in the plane of the meridian (meridian radius of curvature M) and in the plane of the prime vertical, perpendicular to the meridian plane (radius of curvature in the prime vertical N).

The curvature of the meridian $Z = Z(P)$ is

$$\frac{1}{M} = -\frac{d^2Z/d^2P}{[\sqrt{1 + (dZ/dP)^2}]^3} \tag{6.19}$$

Substituting (6.17) and the second derivative obtained by considering (6.18) into (6.19) yields the meridian radius of curvature

$$M = \frac{a(1-e^2)}{(\sqrt{1-e^2\sin^2 B})^3} \tag{6.20}$$

The plane of a parallel circle (oblique section of the rotational ellipsoid) and the vertical plane in the same tangential direction intersect in P at the angle B. The theorem of Meusnier then provides the radius of curvature in the prime vertical:

$$N = P/\cos B \tag{6.21}$$

Using (6.18), one obtains after some manipulations

$$N = a/\sqrt{1-e^2\sin^2 B} \tag{6.22}$$

A comparison of (6.20) and (6.22) shows that $N \geqslant M$. At the poles ($B = \pm 90°$), the polar radius of curvature becomes

$$c = M_{90} = N_{90} = a^2/b \tag{6.23}$$

At the equator ($B = 0°$), there is

$$M_0 = b^2/a, \ N_0 = a \tag{6.24}$$

The curvature of an arbitrary normal section at an azimuth A is computed according **Euler's formula** by

$$\frac{1}{R_A} = \frac{\cos^2 A}{M} + \frac{\sin^2 A}{N} \tag{6.25}$$

Here, R_A is the radius of curvature. The geodetic azimuth A is defined as the angle measured in the horizontal plane between the ellipsoidal meridian plane of P_1 and the vertical plane determined by the normal to P_1 and by the point P_2; A is reckoned from north in the clockwise direction.

6.2.2 Mean Radius of the Spheroid at a Given Point

We have the radius of curvature of a normal section in given latitude B for any azimuth A, namely

$$R_A = MN/(N\cos^2 A + M\sin^2 A) \tag{6.26}$$

To find the mean value of R_A about a point in latitude B, we make use of the theorem of the mean for a function. The theorem is easily demonstrated by means of Fig. 6.3. The slope of the tangent to the curve $y = f(x)$ is given by $f'(x)$, where the prime denotes differentiation, and at the point $Q(\xi, f(\xi))$, the slope is $f'(\xi)$. The slope of the chord PS is $(f(d) - f(c))/(d-c)$ and there exists a point Q as shown such that the slope of the tangent at Q is equal to the slope of the chord PS where $c < \xi < d$, that is, a value can be found such that

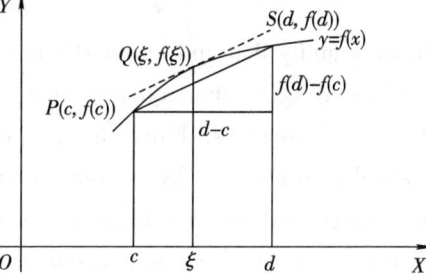

Fig. 6.3 The mean value of a function

$$f'(\xi) = (f(d) - f(c))/(d-c) \tag{6.27}$$

At such a point $f(\xi)$ is defined as the mean value of the function $f(x)$, $c \leqslant x \leqslant d$.

By definition of the definite integral we have

$$\int_c^d f'(x) \, dx = f(d) - f(c) \qquad (6.28)$$

With this value of $f(d) - f(c)$ placed in (6.27) we have

$$f'(\xi) = \frac{1}{d-c} \int_c^d f'(x) \, dx \qquad (6.29)$$

which allows us to compute the mean value of the function $f(x)$, $c \leq x \leq d$ by evaluating the definite integral of this function with limits c and d (The prime denoting differentiation may be omitted in equation (6.29)).

Denoting $f(\xi)$ by R and the limits of A by $c = 0$, $d = 2\pi$ we have from (6.26) and (6.29)

$$\begin{aligned}
R &= \frac{1}{2\pi} \int_0^{2\pi} f(A) \, dA \\
&= \frac{1}{2\pi} \int_0^{2\pi} \frac{MN}{N\cos^2 A + M\sin^2 A} \, dA \\
&= \frac{2}{\pi} \int_0^{\frac{\pi}{2}} \frac{M\sec^2 A}{1 + (M/N)\tan^2 A} \, dA \qquad (6.30) \\
&= \frac{2}{\pi} \sqrt{MN} \left[\arctan\left(\sqrt{\frac{M}{N}} \tan A \right) \right]_0^{\frac{\pi}{2}} \\
&= \frac{2}{\pi} \sqrt{MN} \left(\frac{\pi}{2} - 0 \right) \\
&= \sqrt{MN}
\end{aligned}$$

Thus the mean value of the radius of a spheroid at one of its points is the geometric mean of the principal radii of curvature at the given point.

6.2.3 The Length of the Meridian Arc

A linear quantity that appears mostly in formulas for map projection is the length along the meridian from E (starting at the equator) to A. There is no finite formula for this in terms of simple mathematical functions. From the arc element of the meridian $dX = MdB$ the length X can theoretically be obtained by an integration process; in practice, this integral must be expressed as an infinite series, and we take advantage of the fact that for any spheroid resembling the shape of the earth the value of e^2 is only about $1/150$. To get the formula for X the expression for M is expanded to:

$$a(1 - e^2)\left(1 + \frac{3}{2}e^2 \sin^2 B + \frac{15}{8}e^4 \sin^4 B + \frac{35}{16}e^6 \sin^6 B + \ldots\right) \qquad (6.31)$$

and the powers of $\sin B$ are replaced by the trigonometrical identities:

6 Reference Ellipsoid and Geodetic Coordinate System

$$\sin^2 B = \frac{1}{2} - \frac{1}{2}\cos 2B$$

$$\sin^4 B = \frac{3}{8} - \frac{1}{2}\cos 2B + \frac{1}{8}\cos 4B \qquad (6.32)$$

$$\sin^6 B = \frac{5}{16} - \frac{15}{32}\cos 2B + \frac{3}{16}\cos 4B - \frac{1}{32}\cos 6B$$

$$\vdots$$

Then we find that

$$M = a\left(1 - \frac{1}{4}e^2 - \frac{3}{64}e^4 - \frac{5}{256}e^6 \ldots\right) - a\left(\frac{3}{4}e^2 + \frac{3}{16}e^4 + \frac{45}{512}e^6 \ldots\right)\cos 2B +$$

$$a\left(\frac{15}{64}e^4 + \frac{45}{256}e^6 \ldots\right)\cos 4B - a\left(\frac{35}{512}e^6 \ldots\right)\cos 6B \ldots$$

$$(6.33)$$

and the integration gives:

$$X = a\left(1 - \frac{1}{4}e^2 - \frac{3}{64}e^4 - \frac{5}{256}e^6 \ldots\right)B - a\left(\frac{3}{8}e^2 + \frac{3}{32}e^4 + \frac{45}{1024}e^6 \ldots\right)\sin 2B +$$

$$a\left(\frac{15}{256}e^4 + \frac{45}{1024}e^6 \ldots\right)\sin 4B - a\left(\frac{35}{3072}e^6 \ldots\right)\sin 6B \ldots$$

$$(6.34)$$

in the first term of which the B must be in radian measure. If we put $B = \pi/2$, all trigonometrical terms are zero and we are left with the term in B itself.

This gives the length of the quarter-ellipse from E to P, and it is interesting to recall that the original definition of the meter was a ten-millionth of this length: a recent estimate of the dimensions of an earth-spheroid makes the quadrant 10 002 001 m.

6.2.4 The Length of the Arc of a Circle of Latitude

The length of the arc of a circle of latitude between the geographic longitudes L_1 and L_2 is given according to $dS = N\cos B\, dL$ by

$$S = \int_{L_1}^{L_2} N\cos B\, dL = N\cos B(L_2 - L_1) \qquad (6.35)$$

6.3 Geodesics

6.3.1 Normal Section and Relative Normal Section Line

The line between two points on the surface of an ellipsoid formed by the intersection of the ellipsoid and a plane containing the normal at one point and other point. For example, between points A and B, the normal section at A would contain the normal at A and would pass through point B; the normal section at B would contain the normal at B and would pass through point A. Also called plane curve, curve of normal section, and normal section line.

On the ellipsoid, the geodetic normal sections originating at points A and B, and terminating at

B and A respectively, do not coincide.

Thus at any of the points there are two intersecting geodetic normal sections which in fact may lead to some confusion as to which normal section the azimuth should be measured. To avoid the possibility of confusion, the geodesic line is introduced.

6.3.2 Geodesic

The shortest line between two points on a mathematical surface. The **geodesic** is a line of double curvature and, if the surface is an ellipsoid, lies between the two normal sections. Also termed geodesic line. The term geodesic line is also used when the surface being considered is the reference ellipsoid.

* * * * * * * * * * *

In order to carry out computations on the rotational ellipsoid, the points on the ellipsoid must be connected to one another by surface curves. We consider primarily the normal section (arc) and the geodesic.

The **normal or vertical section** is defined by the curve of intersection of the vertical plane and the ellipsoid. Hence, the direction, which are observed by the theodolite and reduced to the ellipsoid, form angles between normal sections; spatial distances can also be reduced to lengths of normal sections. Since the surface normals of two points on the ellipsoid are in general skewed to each other, the reciprocal normal sections from P_1 to P_2 and from P_2 to P_1 do not coincide (Fig. 6.4). In order to obtain well defined computations, the difference in azimuth $A'_1 - A''_1$ must be taken into account; for $S = 50$ km, this difference amounts to at most only $0.02''$.

Usually, because of its favorable properties in differential geometry, a unique surface curve, the geodesic, is introduced. The line is the shortest connection on the ellipsoid between two points and extends generally between the two reciprocal normal sections (Fig. 6.4).

From the relations

$$\frac{dB}{dS} = \frac{\cos A}{M}, \quad \frac{dL}{dS} = \frac{\sin A}{N \cos B} \tag{6.36}$$

which are taken directly from Fig. 6.5 and according to Clairaut's equation, we have

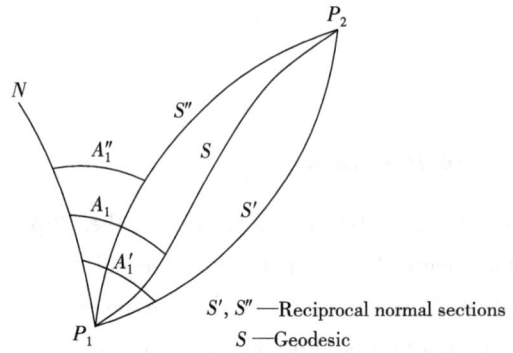

Fig. 6.4 Normal section and geodesic

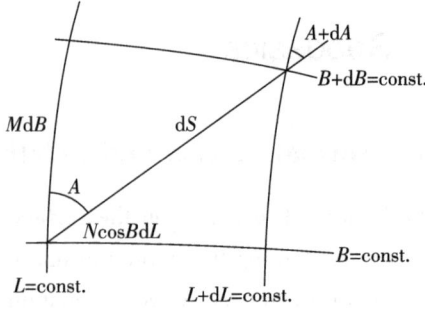

Fig. 6.5 Arc Element in the system of the geographic coordinates

6 Reference Ellipsoid and Geodetic Coordinate System

$$NcosBsinA = \text{const.} \quad (6.37)$$

If one differentiates (6.37) with respect to S, then with (6.36), we have

$$\frac{dA}{dS} = \frac{sinA\tan B}{N} \quad (6.38)$$

For series expansions, (6.36) and (6.38) form an important system of first-order differential equations for the geodesic.

The azimuth A_1' and arc length S' which refer to the normal section are reduced to the azimuth A_1 and the arc length S of the geodesic as follows:

$$A_1' - A_1 = \frac{e^2}{12a^2}\cos^2 B_1 S^2 \sin 2A_1 + \ldots \quad (6.39)$$

$$S' - S = \frac{e^4}{360a^4}\cos^4 B_1 S^5 \sin^2 2B_1 + \ldots \quad (6.40)$$

For $S = 50(200)$ km, the reduction in azimuth amounts to at most $0.007''$ ($0.112''$), and the reduction in arc length is no more than 2×10^{-11} (2×10^{-8}) m. The latter can therefore always be neglected. The azimuth reduction attains the accuracy of first-order angle measurements only for very long lines. C. F. Gauss thoroughly studied the geodesic; its geodetic significance is treated in text books and manuals of geodetic surveying.

6.4 Reductions to the Ellipsoid

6.4.1 Reduction of Horizontal Directions to the Ellipsoid

Directions and distances measured on the surface of the earth need to be reduced to their corresponding values on the reference ellipsoid. The corrections that need to be applied to a direction are for skewness of normals, normal section-geodesic separation, and the deflection of the vertical. The **correction for skewness of normals** is given by:

$$\delta_2 = \frac{e^2}{2b}h_2\cos^2 B_1 \sin 2A_1 \quad (6.41)$$

The **correction for normal section – geodesic separation** is given by:

$$\delta_3 = -\frac{e^2}{12a^2}\cos^2 B_1 \sin 2A_1 \quad (6.42)$$

The **correction for the deflection of the vertical** is given by:

$$\delta_1 = -(\xi \sin A_1 - \eta \cos A_1)\cot Z_1 \quad (6.43)$$

where h_2 is the ellipsoidal height at the target station; B_1 is the geodetic latitude at the observing station; A_1 is the geodetic azimuth of the line from observing station to target station; ξ and η are the respective components of the astrogeodetic deflection of the vertical in the meridian and prime vertical planes at the observing station; Z_1 is the zenith angle at the observing station.

6.4.2 Reduction of Spatial Distances

Electronic measurement of distance yields straight spatial distances l between two points A and B

(Fig. 6.6). These distances may either be used directly for computations in the geodetic coordinate system B, L, h, as in "three-dimensional geodesy", or they may be reduced to the surface of the ellipsoid to obtain chord distances l_0 or geodesic distances s_0.

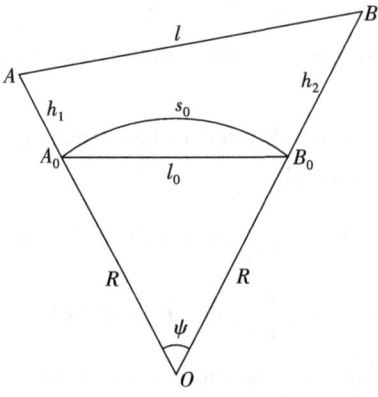

Fig. 6.6 Reduction of spatial distances

We shall again approximate the ellipsoid arc A_0B_0 by a circle arc of radius R that is the mean ellipsoidal radius of curvature along A_0B_0. By applying the law of cosines to the triangle OAB we find

$$l^2 = (R + h_1)^2 + (R + h_2)^2 - 2(R + h_1)(R + h_2)\cos\psi \tag{6.44}$$

with

$$\cos\psi = 1 - 2\sin^2\frac{\psi}{2} \tag{6.45}$$

this is transformed into

$$l^2 = (h_2 - h_1)^2 + 4R^2\left(1 + \frac{h_1}{R}\right)\left(1 + \frac{h_2}{R}\right)\sin^2\frac{\psi}{2} \tag{6.46}$$

and with

$$l_0 = 2R\sin\frac{\psi}{2} \tag{6.47}$$

and the abbreviation $\Delta h = h_2 - h_1$, we obtain

$$l^2 = \Delta h^2 + \left(1 + \frac{h_1}{R}\right)\left(1 + \frac{h_2}{R}\right)l_0^2 \tag{6.48}$$

Hence the chord l_0 and the arc s_0 are expressed by

$$l_0 = \sqrt{\frac{l^2 - \Delta h^2}{(1 + h_1/R)(1 + h_2/R)}} \tag{6.49}$$

$$s_0 = R\psi = 2R\arcsin\frac{l_0}{2R}$$

Ellipsoidal refinements of these formulas may be found in Rinner (Rinner, K. 1956. Ueber die Reduktion grosser elektronisch gemessener Entfernungen. Zeitschrift Fuer Vermessungswesen, V. 81)

6.4.3 Reduction of Astronomical Observations to the Ellipsoid

Now we shall establish the relation between the natural coordinates φ, λ, H and the geodetic

6 Reference Ellipsoid and Geodetic Coordinate System

coordinates B, L, h referring to an ellipsoid according to Helmert's projection. Leaving aside, for the moment, the heights h and H, we may also formulate the problem as the reduction of the astronomical coordinates φ and λ to the ellipsoid. If we also include astronomical observation of azimuth we have to reduce the astronomical coordinates φ and λ and the astronomical azimuth α to the ellipsoid in order to obtain the geodetic coordinates B and L and the geodetic azimuth A. Consider a unit sphere (sphere of radius 1) with its center at the observation station P. The actual plumb line intersects this sphere at the astronomical zenith Z_a, whereas the ellipsoidal normal intersects it at the geodetic zenith Z_g. Fig. 6.7 shows this unit sphere as viewed from above. The line of the sight to the target, for which the azimuth α is measured, intersects the unit sphere at the point T and has the zenith distance Z' and Z with respect to the zeniths Z_a and Z_g. The point P_N corresponds to the direction to the North Pole, which has the zenith distance $90° - \varphi$ and $90° - B$, with respect to Z_a and Z_g; the angle at P_N is the difference $\lambda - L$ between the astronomical and geodetic longitude. The angle at Z_a is the astronomical azimuth α, corresponding to the geodetic azimuth A at Z_g. The point F lies on the astronomical meridian, the great circle connecting P_N and Z_a, so that the angle $Z_a F Z_g$ is $90°$; $\xi = Z_a F$ and $\eta = Z_g F$ are the components of the deflection of the vertical. Consider first the rectangular spherical triangle with corners Z_g, F and P_N. By Napier's rules we have

$$\sin B = \cos(90° - \varphi + \xi)\cos\eta \qquad (6.50)$$
$$\sin\eta = \cos(90° - (\lambda - L))\cos B$$

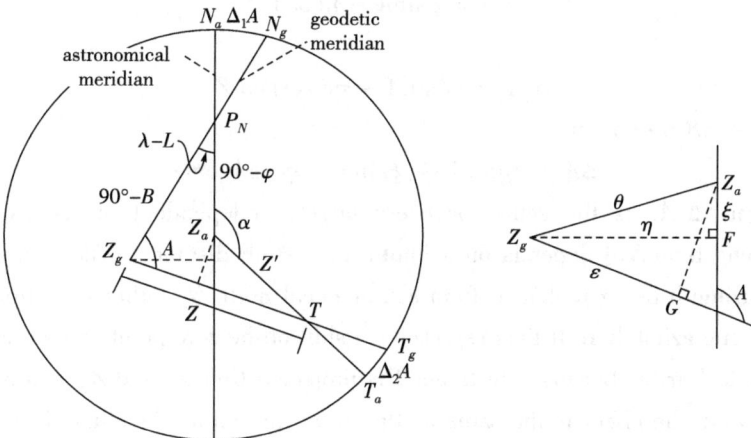

Fig. 6.7 The unit sphere illustrating the deflection of the vertical, as seen from above, and an enlarged view of the central portion of the figure.

For the small angles η and $\lambda - L$ we may use the approximations

$$\cos\eta \doteq 1, \quad \sin\eta \doteq \eta, \quad \cos(90° - (\lambda - L)) = \sin(\lambda - L) \doteq (\lambda - L) \qquad (6.51)$$

Thus we find

$$\xi = \varphi - B \qquad (6.52)$$
$$\eta = (\lambda - L)\cos B$$

These are the **basic equations** that express the components ξ and η of the deflection of the vertical in terms of the geographic (astronomical and geodetic) coordinates, thus linking astronomical and geodetic coordinates.

The difference in azimuth
$$\Delta A = \alpha - A \tag{6.53}$$
consists of two parts, $\Delta_1 A$ and $\Delta_2 A$ (Fig. 6.7):
$$\Delta A = \Delta_1 A + \Delta_2 A \tag{6.54}$$
$\Delta_1 A$ is obtained from the spherical triangle $N_g N_a P_N$, which is obviously similar to the triangle $Z_g F P_N$ previously used, $N_g P_N = B$ corresponding to $Z_g P_N = 90° - B$, and $\Delta_1 A$ corresponding to η. The equation corresponding to $\eta = (\lambda - L)\cos B$ is thus
$$\Delta_1 A = (\lambda - L)\sin B \tag{6.55}$$
together with $\eta = (\lambda - L)\cos B$ this becomes
$$\Delta_1 A = \eta \tan B \tag{6.56}$$

On introducing a point G on the great circle connecting Z_g and T so that the angle $Z_a G Z_g$ is 90°, and putting $Z_a G = \delta$, we see that the figure $Z_a G T T_g T_a$ has the same geometry as the figure $Z_g F P_N N_a N_g$, so that $\Delta_2 A$, δ, Z' correspond to $\Delta_1 A$, η, $90° - B$. The equation corresponding to $\Delta_1 A = \eta \tan B$ is thus
$$\Delta_2 A = \delta \cot Z' = \delta \cot Z \tag{6.57}$$

Since the small figure $Z_a F Z_g G$ may be considered plane (see the enlarged section of Fig. 6.7), we get by the usual formula of the transformation of plane coordinates
$$\delta = \xi \sin A - \eta \cos A \tag{6.58}$$
so that
$$\Delta_2 A = (\xi \sin A - \eta \cos A)\cot Z \tag{6.59}$$
and with $\Delta_1 A = \eta \tan B$ we obtain
$$\Delta A = \eta \tan B + (\xi \sin A - \eta \cos A)\cot Z \tag{6.60}$$

The first term, $\Delta_1 A$, is the same for every target, independent of its azimuth and zenith distance; the second term $\Delta_2 A$ depends on azimuth and zenith distance. The term $\Delta_1 A$ results from the astronomical azimuth being reckoned from astronomical north N_a rather than from geodetic north N_g, as is the geodetic azimuth A. It thus represents a shift of the zero point, which is the same for all targets. The term $\Delta_2 A$ arises because the target T is projected from Z_a and Z_g onto different points T_a and T_g of the horizon; the effect is the same as that of an inaccurate leveling of the theodolite.

Usually in first-order triangulation the lines of sight are almost horizontal, so that $Z = 90°$, cot $Z = 0$. Therefore, the correction $\Delta_2 A$ can in general be neglected and we thus get
$$\Delta A = \eta \tan B = (\lambda - L)\sin B \tag{6.61}$$

This is Laplace's equation in its usual simplified form. It is remarkable that the differences $\Delta A = \alpha - A$ and $\lambda - L$ should be related in such a simple way.

For later reference we note that total deflection of the vertical—that is, the angle θ between the actual plumb line and the ellipsoid normal—is given by
$$\theta = \sqrt{\xi^2 + \eta^2} \tag{6.62}$$
and that the deflection component ε in the direction of the azimuth A is
$$\varepsilon = \xi \cos A + \eta \sin A \tag{6.63}$$

Both relations can be read immediately from the enlarged section of Fig. 6.7; δ and ε are related to ξ and η by a plane coordinate transformation.

Finally, the relationship between the orthometric height H above the geoid and the geometric height h above the ellipsoid can be written down immediately because from Fig. 6.1 (cf. "Projection Method") we read that, to a sufficient approximation,

$$h = H + N \tag{6.64}$$

Thus the conversion formulas from natural to geodetic coordinates are

$$\begin{aligned} B &= \varphi - \xi \\ L &= \lambda - \eta \sec B \\ h &= H + N \end{aligned} \tag{6.65}$$

and the corresponding formula for the azimuth is

$$A = \alpha - \eta \tan B \tag{6.66}$$

In the application of these formulas we need the geoidal undulation N and the deflection components ξ and η with respect to the reference ellipsoid used.

Two points should be noted:

(1) The axis of the reference ellipsoid is parallel to the earth's axis of rotation (otherwise there would be two different poles P_N in Fig. 6.7), but it need not be in an absolute position, its center coinciding with the earth's center of gravity.

(2) The deflection components ξ and η refer directly to the point on the ground at which the astronomical observations are made, and not to the geoid.

If components ξ and η of the deflection of the vertical are computed gravimetrically for the geoid by Vening Meinesz' formula, then B, L, h, and A refer to an ellipsoid in absolute position, but care should be taken because of the curvature of the plumb line.

It should also be mentioned that the ellipsoidal azimuth A refers to the actual target T, which does not in general lie on the ellipsoid. For the conventional method of computation on the ellipsoid one wishes the azimuth to refer to a target T_0 on the ellipsoid, which is the point at the foot of the normal through T. Furthermore, A refers to what is called a normal section of the ellipsoid, rather than to a geodesic line, which is used in computation. In either case very small azimuth reductions are necessary; since these reductions are purely problems in ellipsoidal geometry, the reader is referred to any textbook on geometrical geodesy.

6.4.4 Laplace Equation (Azimuth) and Its Role

An expression relating astronomic and geodetic azimuths with astronomic and geodetic longitudes and latitudes:

$$\alpha - A = (\lambda - L)\sin\varphi \tag{6.67}$$

where α is astronomic azimuth; A is geodetic azimuth; λ is astronomic longitude; L is geodetic longitude; and φ is latitude (azimuth clockwise from north, longitudes positive to the east).

This equation is very important in the strength of a triangulation net because it provides a means of controlling angular distortions through the net. Triangulation stations at which the Laplace equation can be formed (i. e., astro-longitude, and azimuth were observed) are known as Laplace stations, or Laplace points.

Assume that at a point P on the earth's surface, the astronomical azimuth α and the zenith distance Z' to another point Q has been measured. Both α and Z' refer to the actual plumb line at P, since the vertical axis of the theodolite is made to coincide with this plumb line. If, instead, the theodolite axis could be made to coincide with the normal, at P, to the reference ellipsoid, then we would measure a "geodetic" azimuth A and a "geodetic" zenith distance Z. Astronomical azimuth α and astronomical zenith distance Z' are related to their geodetic counterparts by

$$\alpha - A = \eta \tan B + (\xi \sin A - \eta \cos A) \cot Z \qquad (6.68)$$

$$Z' - Z = -(\xi \cos A + \eta \sin A) \qquad (6.69)$$

Especially in first-order triangulation, all lines of sight will usually be almost horizontal, that is, $Z = 90°$. Then (6.68) reduces to

$$\alpha - A = \eta \tan B \qquad (6.70)$$

On using the equation ($\eta = (\lambda - L) \cos B$), this becomes

$$\alpha - A = (\lambda - L) \sin B \qquad (6.71)$$

Equation (6.71) is called Laplace's equation, or Laplace condition. If it is possible, at a station, to observe astronomical both longitude λ and azimuth α, and if their geodetic counterparts L and A can be found independently (by the computation of the triangulation), then (6.71) forms a condition which the four quantities α, A, λ, L must satisfy.

It is clear that, if Z cannot be put $90°$, then the exact form of the Laplace equation, by $\xi = \varphi - B$, $\eta = (\lambda - L) \cos B$ and (6.71), is

$$\alpha - A = (\lambda - L) \sin B + [(\varphi - B) \sin A - (\lambda - L) \cos B \cos A] \cot Z \qquad (6.72)$$

All these equations have been derived on the basis of parallelism of the ellipsoidal axes xyz and the global axes XYZ. Their use in the computation of a triangulation will, therefore, serve to ensure the parallelism of these two rectangular frames.

To understand the situation, let us briefly consider, from a geometric point of view, how a triangulation is computed. To make the geometric structure transparent, **we presuppose errorless measurements.**

Let us first assume, for the sake of simplicity, that all stations of the triangulation lie on the surface of the reference ellipsoid (as a second step, we shall rid ourselves of this oversimplification). However, the plumb lines are not taken to coincide with the ellipsoidal normal; that is, deflections of the vertical ξ, η are admitted. Distances (straight spatial distances, i. e., chords between ellipsoidal points) and horizontal angles are given to the extent necessary to determine the geometric configuration on the ellipsoid. Astronomical coordinates φ and λ have been observed at each station. At one station, the geodetic origin P_0, an astronomical azimuth α_0, has been measured in addition to φ_0 and λ_0. For P_0, also geocentric coordinates X_0, Y_0, Z_0 are given.

The geodetic coordinates B and L of the triangulation points (h is zero by our assumption) can be obtained as follows. According to the transformation equation from the (B, L, h) to the (X, Y, Z), for the geodetic origin P_0, we have

6 Reference Ellipsoid and Geodetic Coordinate System

$$\underline{X}_0 = \underline{X}(B_0, L_0, h_0) \tag{6.73}$$

These equations are solved for B_0, L_0, h_0; since the geodetic origin P_0 is on the ellipsoid, there must be $h_0 = 0$ with our errorless data.

With the measured astronomical coordinates φ_0 and λ_0, we can compute deflections of the vertical at P_0:

$$\xi_0 = \varphi_0 - B_0, \quad \eta_0 = (\lambda_0 - L_0)\cos B_0 \tag{6.74}$$

Then (6.68), which practically reduces to (6.70), gives

$$A_{01} = \alpha_{01} - \eta_0 \tan B_0 \tag{6.75}$$

determining the initial ellipsoidal azimuth A_{01} from the measured astronomical azimuth α_{01}.

Horizontal angles measured at P_0 can also be reduced to the ellipsoid:

$$\gamma_{12}^{\text{ell.}} = \gamma_{12}^{\text{meas.}} + (\xi_0 \sin A_{01} - \eta_0 \cos A_{01})\cot Z_{01} - (\xi_0 \sin A_{02} - \eta_0 \cos A_{02})\cot Z_{02} \tag{6.76}$$

by taking the difference of two equations (6.68). The main term $\eta \tan B$ has dropped out, so that for nearly horizontal lines of sight the reduction of horizontal angles will be very small and often negligible.

Now the azimuths of all directions P_0P_1, P_0P_2, ... initiating from P_0 can be computed (Fig. 6.8):

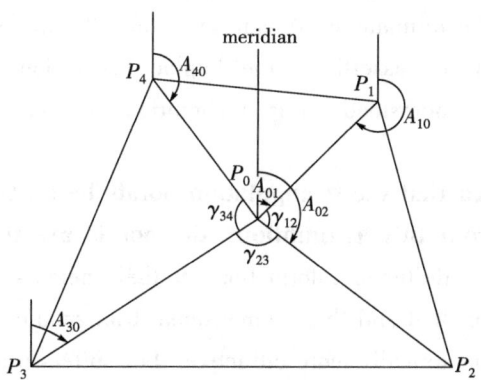

Fig. 6.8 Computation of triangulation net

$$A_{01} \text{ by} \tag{6.75}$$
$$A_{02} = A_{01} + \gamma_{12}$$
$$A_{03} = A_{02} + \gamma_{23} \tag{6.77}$$

where γ_{ij} denote ellipsoidal values $\gamma_{ij}^{\text{ell.}}$.

By well-known ellipsoidal computation methods using chords we can then compute the geodetic coordinates of the triangulation points surrounding P_0 (in the figure: P_1, P_2, P_3, P_4) and finally the reverse azimuths $A_{10}, A_{20}, A_{30}, A_{40}$ (without any additional azimuth measurements!).

The next point, say P_1, can be treated in exactly the same way. The previous ellipsoidal computations have given B_1, L_1, A_{10}; from astronomical observations we know φ_1, λ_1, so that gives ξ_1, η_1. Then the horizontal angles measured in P_1 can be reduced to the ellipsoid (if necessary) and we can proceed to the points surrounding P_1, and so on.

In this way we obtain geodetic coordinates B, L of all stations and geodetic azimuths A of all sides.

The essential point is that only one astronomically measured azimuth (in our case α_{01}) is needed to completely fix the orientation of the triangulation, the parallelism of the ellipsoidal and terrestrial Z-axis is ensured by the choice of the geodetic datum.

What happens if another astronomical azimuth, say α_{56}, is measured? Since φ_5, λ_5, B_5, L_5 are known according to the foregoing considerations, we get ξ_5 and η_5 and can thus compute A_{56} by (6.68). However, we already know A_{56} from the preceding ellipsoidal computations. Thus the two values for A_{56}, one from ellipsoidal computations and one from reduction of the measured α_{56} to the ellipsoid, must coincide. In other terms, the Laplace condition (6.71) must be satisfied

$$(\alpha_{56} - A_{56}) = (\lambda_5 - L_5)\sin B_5 \qquad (6.78)$$

A_{56} being the value from ellipsoidal computations (more exact is of course, (6.72)).

If our observations are errorless, as we have assumed, then this Laplace equation will automatically be satisfied because the geometric situation is uniquely determined. **If the observations are affected by measuring errors**, then the Laplace condition will no longer be satisfied, but can be enforced by an adjustment. Any additional astronomical azimuth measurement gives another Laplace condition for adjustment. In this way the ideal geometrical condition (exact orientation and parallelism of coordinate axes) can be restored to the best possible extent.

It is in this sense that the assertion "the Laplace condition ensures net orientation and parallelism of axes" is to be understood: it is a condition, not for fixing the geometry, but for adjusting measuring errors.

So far we have assumed that the triangulation points lie on the surface of ellipsoid. We shall now free ourselves from this assumption: the points are now situated on the earth's surface. Therefore, we need additional information on their heights. Two principal methods are available: the astrogeodetic method and three dimensional triangulation using zenith distances.

The second method is theoretically more attractive, but suffers practical difficulties with zenith distances. So the first method is preferred in practice.

In the astrogeodetic method, geoid height determination by astronomic leveling, combined with orthometric heights, is used to reduce this case to the first simple case of points on the ellipsoid by an iterative procedure. The geometrical situation with respect to the Laplace equation remains essentially the same.

In three-dimensional triangulation, we may reduce the observations (azimuths, horizontal angles, zenith distances) to the ellipsoidal normal and use geodetic coordinates. In this way, the situation is again brought back to the first case, the Helmert projections of the triangulation points onto the ellipsoid playing the role of the former points situated on the ellipsoid. Again the geometric situation with respect to the Laplace azimuth equation remains the same. One astronomical azimuth is needed to fix the geometric orientation, other azimuths help to adjust for measuring errors and thus to produce a geometry that approaches as well as possible the ideal case: proper orientation and parallelism.

If, in this case, more zenith distances are measured as it is geometrically necessary, then the surplus measurements provide conditions of the form (6.69). In this sense, it is understandable that the system of the two equations (6.68) and (6.69) has recently been called "extended Laplace equation". However, such conditions on zenith distances only help to improve heights; they have practically no effect on orientation and axes. Therefore, the present author prefers to reserve the name "Laplace equation" to the azimuth condition.

It has sometimes been asked how many Laplace conditions are necessary to determine the geometric orientation. In the author's opinion, this question should not be put in this way, because it is ambiguous: the answer may run from zero (if only surplus conditions are considered) to any number (if the reduction of any horizontal or vertical angle measurement is viewed as an application of the "extended Laplace equation" (6.68) and (6.69)). It seems clearer to ask for the number of measurements necessary to fix the geometry: φ and λ at each point, enough horizontal angles, distances and (as the case may be) zenith distances to determine the geometrical configuration, and one astronomical azimuth to fix orientation. A total number of n astronomical azimuths (at n different points) gives $(n-1)$ Laplace conditions, geometrically superfluous but practically of the greatest importance. The more and the better distributed these azimuths are, the better will be the orientation of the network and of the coordinate axes.

6.4.5 Solution of Ellipsoidal Triangles

For the determination of geographic coordinates from distance and azimuth, the unobserved sides and angles must be computed using ellipsoidal triangles. For the lengths of sides in a first-order triangulation, the ellipsoidal computations can be replaced by calculations on a sphere. In this respect, the ellipsoid is approximated by the Gaussian osculating sphere tangent at $P_0(B_0)$ and having a radius

$$R_0 = \sqrt{M_0 N_0} \qquad (6.79)$$

The latitude of tangency B_0 can be chosen as the arithmetic mean of the geographic latitudes of the vertices of the particular triangle. Within a radius of 150 km about the point of tangency of the Gaussian osculating sphere, the errors in direction caused by the spherical approximation remains less than $0.005''$.

For the solution of a spherical triangle having sides a, b, c and angles A, B, C, the spherical law of sines yields

$$\frac{\sin A}{\sin B} = \frac{\sin(a/R)}{\sin(b/R)} \qquad (6.80)$$

Expanding $\sin(a/R)$ and $\sin(b/R)$ in series up to $O(1/R^3)$, one obtains relations corresponding to the law of sines for the plane. If the spherical angles are reduced by one third of the spherical excess, then this gives Legendre's equation:

$$\frac{\sin(A - E/3)}{\sin(B - E/3)} = \frac{a}{b} \qquad (6.81)$$

If the spherical sides are given increments (additaments), then the result is the Soldner method of additaments:

$$\frac{\sin A}{\sin B} = \frac{a - a^3/6R^2}{b - b^3/6R^2} \tag{6.82}$$

The spherical excess E appearing in (6.81) is the surplus over 180° of the angle sum of a spherical triangle. It is computed according to

$$E = F/R^2 \tag{6.83}$$

where F is the area of the plane triangle computed from the lengths of the spherical sides. For an equilateral triangle with $S = 50$ km, $E = 5.48''$.

6.4.6 Legendre's Theorem

One of the operations that will often have to be carried out in survey computation is the solution of triangles; this involves the relationships between the sides and angles of triangles.

Surveyors are concerned with the linear lengths of sides of triangles, rather than with angles they may represent at the centre of some sphere. In the early days of geodetic computation, triangles were treated as spherical, the linear lengths of the sides were converted to angle units using a reasonable radius for the sphere, and sides computed in angle measure were converted back to linear values: obviously a tedious process! It was Legendre who provided another solution, published as an exercise in his book on trigonometry in 1813.

Legendre's Theorem states, in effect, that if a spherical triangle that is small in relation to the size of the sphere is regarded as made of three pieces of string, and is removed from the sphere and stretched out flat, the resulting plane triangle will have angles that differ by almost exactly equal amounts from the spherical angles: the differences must therefore each be $1/3E$. This means that if the linear length of one side of a spherical triangle is known, and the angles are each reduced by $1/3E$, then the relationships

$$\frac{a}{\sin\left(A - \frac{1}{3}E\right)} = \frac{b}{\sin\left(B - \frac{1}{3}E\right)} = \frac{c}{\sin\left(C - \frac{1}{3}E\right)} \tag{6.84}$$

are so nearly true that they can be used for calculating the lengths of the other two sides in triangles of geodetic dimensions.

A proof of the theorem is outlined below; a different proof, based on the idea of stretching out the triangle, will be found elsewhere. The spherical triangle in Fig. 6.9(1) has sides expressed as usual in angle measure, α, β, γ: their lengths will be $\alpha R, \beta R, \gamma R$, if R is the radius of the sphere. Consider a plane triangle, Fig. 6.9(2), having angles as shown, which of course add up to the correct 180°. We have

$$\frac{x}{y} = \frac{\sin\left(A - \frac{1}{3}E\right)}{\sin\left(B - \frac{1}{3}E\right)} = \frac{\sin A - \frac{1}{3}E\cos A \ldots}{\sin B - \frac{1}{3}E\cos B \ldots} \tag{6.85}$$

Now, for calculating the spherical excess, which is very small, the area of the spherical triangle can be taken as

$$\frac{1}{2}\beta R \gamma R \sin A, \quad \text{or} \quad \frac{1}{2}\gamma R \alpha R \sin B, \quad \text{or} \quad \frac{1}{2}\alpha R \beta R \sin C$$

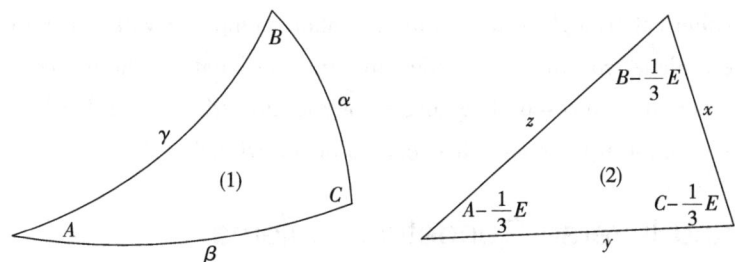

Fig. 6.9 Legendre's theorem

so, using the formula (The spherical excess of the triangle E = (triangle area)/R^2), we have

$$E = \frac{1}{2}\beta\gamma\sin A = \frac{1}{2}\gamma\alpha\sin B \quad \left(= \frac{1}{2}\alpha\beta\sin C, \text{ also}\right) \tag{6.86}$$

By the same token, the cosines which appear in the fraction x/y can be replaced by the formulas

$$\cos A = \frac{\beta^2 + \gamma^2 - \alpha^2}{2\beta\gamma} \text{ and } \cos B = \frac{\gamma^2 + \alpha^2 - \beta^2}{2\gamma\alpha} \tag{6.87}$$

derived from plane trigonometry. Making the substitutions, we find

$$\frac{x}{y} = \frac{\sin A \left[1 - \frac{1}{12}(\beta^2 + \gamma^2 - \alpha^2)\right]}{\sin B \left[1 - \frac{1}{12}(\gamma^2 + \alpha^2 - \beta^2)\right]} \tag{6.88}$$

But

$$\frac{\sin A}{\sin B} = \frac{\sin\alpha}{\sin\beta}(\text{exactly}) = \frac{\alpha - \frac{1}{6}\alpha^3\ldots}{\beta - \frac{1}{6}\beta^3\ldots} = \frac{\alpha\left(1 - \frac{1}{6}\alpha^2\ldots\right)}{\beta\left(1 - \frac{1}{6}\beta^2\ldots\right)} \tag{6.89}$$

Substituting for $\frac{\sin A}{\sin B}$ and retaining only 2nd order terms in the series, we get

$$\frac{x}{y} = \frac{\alpha}{\beta} \cdot \frac{1 - \frac{1}{12}(\beta^2 + \gamma^2 + \alpha^2)\ldots}{1 - \frac{1}{12}(\gamma^2 + \alpha^2 + \beta^2)\ldots} \tag{6.90}$$

To second order, the series cancel out and we are left with $\frac{x}{y} \approx \frac{\alpha}{\beta}$, showing that the sides of the plane triangle are proportional to the lengths of the sides of the spherical triangle, to the order of accuracy mentioned.

If further study is made of the formulas, it can be shown that the discrepancies involved in using Legendre's Theorem are quite amazingly small: if the triangle is big enough to cover the whole of Great Britain, the ratios x/α, y/β and z/γ are equal to within 2 parts per million!

It must be emphasized that the use of Legendre's Theorem is only a computing aid: the Legendre triangle associated with a spherical triangle cannot be shown geometrically related to it by any simple construction.

* * * * * * * * * * *

If we have a spherical triangle whose sides are short compared with the radius of the sphere, and if we also have a plane triangle whose sides are equal in length to the corresponding sides of the spherical triangle, then the corresponding angles of the two triangles differ by approximately the same quantity, this is one-third the spherical excess of the triangle.

6.5 Direct and Inverse Geodetic Problems

6.5.1 Reduced Latitude

The angle at the center of a sphere which is tangent to the spheroid along the geodetic equator, between the plane of the equator and the radius to the point intersected on the sphere by a straight line perpendicular to the plane of the equator and passing through the point on the spheroid whose reduced latitude is defined. Reduced latitude is an auxiliary latitude used in problems of geodesy and cartography. In astronomical work, when the term reduced latitude is used, geocentric latitude is meant. Also called geometric latitude; parametric latitude.

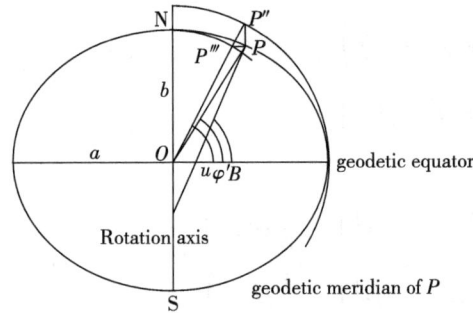

Fig. 6.10 The Meridian Plane

For computational purposes two special latitudes (reduced latitude and geocentric latitude) are introduced in Fig. 6.10 showing the meridian plane of point P. The line OP is the geocentric radius of P. The angle φ', between the radius and the geodetic equator is the geocentric latitude. The angle u between the line OP'' and the equator is called reduced latitude. The point P'' is obtained by projecting P paralled to the rotation axis to the auxiliary circle of radius a. The point P''' is obtained by projecting P raralled to the geodetic equator to the auxiliary circle of radius b. The relations between the three latitudes are:

$$\tan\varphi' = \frac{b}{a}\tan u = \frac{b^2}{a^2}\tan B$$
$$\tan\varphi' = (1-f)\tan u = (1-f)^2 \tan B \quad (6.91)$$
$$\tan\varphi' = \sqrt{1-e^2}\tan u = (1-e^2)\tan B$$

The length ρ of the geocentric radius is given by

$$\rho = \frac{a\sqrt{1-e^2}}{\sqrt{1-e^2\cos^2\varphi'}} \quad (6.92)$$

6.5.2 Direct and Inverse Geodetic Problems

A basic problem in geodetic computation, generally referred to by continental writers as the **principal problem of geodesy,** is that of calculating the spheroid latitude and longitude of a point when its bearing and distance from a point of known coordinates are given. For instance, after a system of major triangulation has been adjusted to be internally consistent, positions of the points on a spheroid of reference need to be calculated, to provide a framework for further surveys or for mapping.

To locate the system on the spheroid, it is sufficient to specify the geodetic coordinates of one point and the bearing of one line; other positions follow by exact computation, using formulas appropriate to the geometry of the spheroid.

To be more specific, a point P_1 has latitude B and longitude L on a spheroid of reference: the bearing at P_1 to another point P_2 is A and the distance from P_1 to P_2 is S. What are the latitude and longitude of P_2 ? There are several well-known sets of formulas for solving this problem. It is necessary to consider what computational precision is required: obviously this must be related to the order of accuracy likely to be achieved in the adjusted survey. If the bearing of a line 40 km long is correct to $(1/2)''$, the relative positional accuracy between the two ends is 0.1 m, or a fractional accuracy of 1/400 000. The accuracy of careful length measurement by EDM approaches this figure. The distance along a meridian equivalent to $1''$ of latitude is about 31 m, so computation to $0.001''$ of latitude corresponds with 0.03 m.

Thus, to avoid accumulation of error in extensive precise computations, length may be taken to 0.01 m, angles to $0.01''$, latitudes and longitudes to $0.0001''$, but there is no point in quoting final values closer than $0.1''$ in angles and bearings, or $0.001''$ for coordinates.

※ ※ ※ ※ ※ ※ ※ ※ ※ ※ ※

In ellipsoidal computations, the following problems repeatedly arise:

(1) To compute the geographic coordinates B_2, L_2 of point P_2, as well as the azimuth A_2, given the coordinates B_1, L_1 of point P_1, the azimuth A_1, and the distance S.

(2) To compute the azimuths A_1, A_2 and the distance S, given the coordinates B_1, L_1, B_2, L_2 of the points P_1, P_2.

These problems are known as the **direct and inverse geodetic problems**, respectively. Due to their importance in the design of triangulation networks, they have been called the first and second principal geodetic problems in the German literature. In either case, one is concerned with the solution of the ellipsoidal polar triangle P_1NP_2 (Fig. 6.11). If the geodesic is introduced as the surface curve between P_1 and P_2, then the problems above demand the integration of the differential equations of the geodesic and the solution for the desired quantities. The numerous solutions can be divided into three groups.

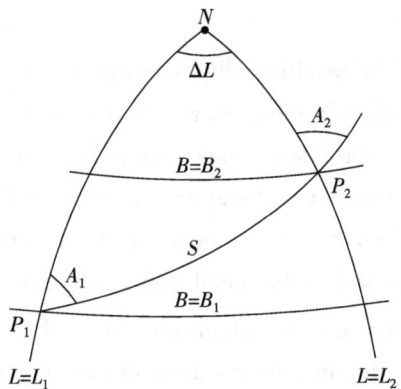

Fig. 6.11 Direct and inverse geodetic problem

The solutions in the first group are based on the integration of the system of differential equations of the geodesic, that is, $dB/dS = \cos A/M$, $dL/dS = \sin A/N\cos B$, and $dA/dS = \sin A\tan B/N$. Here, Legendre (1806) carried out Taylor series expansions of the latitude, longitude, and azimuth differences as functions of arc length:

$$B_2 - B_1 = \left(\frac{dB}{dS}\right)_1 S + \frac{1}{2}\left(\frac{d^2B}{dS^2}\right)_1 S + \ldots$$

$$L_2 - L_1 = \left(\frac{dL}{dS}\right)_1 S + \frac{1}{2}\left(\frac{d^2L}{dS^2}\right)_1 S + \ldots \quad (6.93)$$

$$A_2 - A_1 = \left(\frac{dA}{dS}\right)_1 S + \frac{1}{2}\left(\frac{d^2A}{dS^2}\right)_1 S + \ldots$$

Using the first derivatives, one can also compute the higher derivatives. Since the series are expanded with respect to S, they converge slowly, so that in general, they are applicable only up to a range of 150 km. Dorrer (1966) integrated the differential equations numerically according to the Runge-Kutta method.

In the methods developed by Bessel (1826) and Helmert (1880), the ellipsoidal polar triangle is transferred to a concentric sphere (radius a). The computations are then performed on this sphere and a transformation is subsequently made back to the ellipsoid. The reduced latitude u is used for the spherical latitude; u is related to B by $\tan u = \frac{b}{a}\tan B$. If one transfers the azimuth A_1 to the sphere, then by Clairaut's equation ($N\cos B\sin A$ = const., that is, $\cos u\sin A$ = const.), the ellipsoidal azimuths are preserved in this transference. The relations between the ellipsoidal distance S and spherical distance σ, as well as those between the ellipsoidal and spherical longitude differences ΔL and $\Delta \lambda$ are derived by combining the equations

$$du/d\sigma = \cos A, \quad d\lambda/d\sigma = \sin A/\cos u \quad (6.94)$$

which are valid on the sphere, with the corresponding ellipsoidal formulas $dB/dS = \cos A$ and $dL/dS = \sin A/N\cos B$. After some manipulations, the following differential equations are obtained:

$$dS = a\sqrt{1 - e^2\cos^2 u}\,d\sigma \quad (6.95)$$
$$dL = \sqrt{1 - e^2\cos^2 u}\,d\lambda$$

The resulting elliptic integrals can be solved either by expanding the square roots in series and subsequently integrating term by term, or by numerical methods. This process exhibits favorable convergence since one solves for only small differences between the ellipsoidal and spherical quantities. It is therefore suitable also for computations involving very long geodesics.

Finally, the principal geodetic problems can be solved using a conformal projection to the sphere with subsequent spherical calculations and a transformation back to the ellipsoid. Simple transformation equations are obtained by employing the Gaussian osculating sphere ($R_0 = \sqrt{M_0 N_0}$). Since the projection causes distortions, the azimuths and distances are subject to reductions. This method is also suitable for long lines.

Within limited areas, one can also take advantage of closed formulas for the computation of coordinates.

6.5.3 Solution of the Direct and Inverse Problems on Reference Ellipsoids by Point-by-point Integration using Programmable Pocket Calculators

Two fundamental geodetic calculations on a reference ellipsoid are generally known as the direct (forward) and inverse problems. The direct problem involves computation of the latitude, longitude, and back azimuth of a point whose distance and azimuth from a given geodetic origin are known. The inverse problem calculates the distance and azimuths between two known geographic positions.

To solve the direct and/or the inverse problem on the surface of a reference ellipsoid, there are more than 50 different sets of formulas, usually bearing the name of the developer of the method, from which to choose. If 1 cm accuracy is desired most of these methods cannot be used for distances much greater than 150 km. Some can be used up to 400 km but only a very few from 400 km to 20 000 km, the latter being the maximum possible distance between two points on the earth's surface.

Almost all of these methods start with the same differential formulas for a geodetic line element on the surface. The classical procedure for solution is by series development using either Legendre's, Gauss Mid-Latitude, or Bessel-Helmert series. Variations in these series developments, utilizing substitutions, projections to auxiliary surfaces, integration of the terms of the series, have produced most of the 50 methods mentioned. However, the original differential equations can be directly integrated, and the practical implementation of such a method on a digital computer was published by Kivioja in 1971.

Acknowledging that many users do not have ready access to a computer and recognizing the development and capabilities of hand-held computer-calculators, this paper describes further development of Kivioja's earlier work, enabling a programmable pocket calculator essentially to replace the digital computer. Users of the program need only enter the known data of the geodetic problem: geographic position, distance, and azimuth for the direct problem, or two geographic positions for the inverse problem, plus the semiaxes of the reference ellipsoid for either case. The method was programmed on a Hewlett-Packard HP-67 programmable pocket calculator, although the algorithms developed could be employed on any calculator with storage and program memory equivalent to the HP-67. This program can also be used on the HP-97 and HP-41C calculators.

In the direct problem, the **"point-by-point" method,** as this solution by integration is called, divides the given distance into n equal sections each of length Δs so that $s = n \cdot \Delta s$. The azimuth of each element Δs is controlled by Clairaut's Equation for a geodetic line to "force" each element to lie on the geodetic line. The program computes the latitude and longitude increments for each element and cumulatively adds them to the geographic coordinates of the geodetic origin. As in numerical integrations in general, the accuracy of the result can be increased by decreasing the length of the geodetic line elements Δs.

The inverse problem described here computes approximations for the distance and one azimuth and, starting from one of the known positions, computes the coordinates of the second position by the same algorithm used in the direct problem. The computed coordinates will not generally coincide with the given values, but the differences between the computed and the known points are used to correct the initial approximations for the distance and azimuth.

Computing time of the program depends upon the number of integrations over the geodetic line. It is 5 minutes or less for most problems and all EDM lines. In 11 seconds the HP-67 performs all the calculations associated with one integration over an element Δs. For centimeter accuracy in either geodetic problem, the length of Δs, which be varied by the user, should not exceed 4 km. Very long distances may be calculated with maximum accuracy by overnight runs of the calculator.

Vocabulary

deflection of the vertical /di'flekʃən ɔv ðə 'vəːtikəl/ 垂线偏差
geodesic /ˌdʒiːəu'desik/ adj. 大地测量的, 测量的; n. 大地线
geodetic datum /ˌdʒiːə'detik 'deitəm/ 大地基准
geodetic reference system (GRS) /ˌdʒiːə'detik 'refrəns 'sistəm/ 大地测量参考系
horizontal direction /ˌhɔri'zɔntəl di'rekʃən/ 水平方向
normal section /'nɔːməl 'sekʃən/ 法截面
polar radius of curvature /'pəulə 'reidjəs ɔv 'kəːvəˌtʃuə/ 极曲率半径
prime vertical /praim 'vəːtikəl/ 卯酉圈
rotational ellipsoid /rəu'teiʃən əli'lipsɔid/ 旋转椭球
spherical excess /'sfiərikəl 'ekses/ 球面角盈
spheroid /'sfiərɔid/ n. 球状体, 旋转椭球
zenith distance /'ziːniθ 'distəns/ 天顶距

7 Map Projection

7.1 Introduction of Map Projection

The ultimate object of most surveying operations is mapping. A map is a plane representation of the curved surface of the spheroid, on which positions are defined as latitudes and longitudes. The construction of a map therefore requires the transformation of spheroid coordinates into a plane system: such a transformation is a **map projection.**

It is however unnecessary to find spheroid coordinates of every point surveyed and then transform them to the plane system: the usual procedure is to calculate spheroid latitudes and longitudes of the main framework of a survey; some or all of these values may also be used, in conjunction with astronomical observations, in studies of the shape and size of the earth. The rest of the points are needed only for control of mapping, engineering surveys, etc. Nevertheless, since all the survey measurements must be made on the earth, it is necessary to ensure that all positions are correctly expressed in the plane system. Two problems have now to be considered: the transformation of spheroid coordinates expressed as latitudes and longitudes, into a system of rectangular coordinates for plotting on a plane surface, by using the geometry and mathematics of a map projection; and the methods a surveyor may adopt for obtaining correct plane coordinates of surveyed points directly from his observations on the earth, thus bypassing, as it were, their positions on the surface of reference, but of course taking into account the structure of the projection in which the points have to be located.

There are innumerable constructions which can be used for representing the spheroid surface on a plane, and many of these can be seen in atlases. Because an atlas map must comprise a large portion, or perhaps the whole, of the earth, the distortion of shape and variations of scale unavoidable in any plane representation are quite conspicuous. The surveyor and engineer, on the other hand, are concerned with maps on much larger scales, typically 1/50 000 for topographical maps, or 1/2 000 for cadastral and engineering plans. These will usually be based on a regional or national projection system limited to a very small fraction of the spheroid, such an area, for instance, as that of Great Britain.

If the area covered on one coordinate system is sufficiently small, it is easy to devise a projection structure such that the distortions mentioned above are imperceptible, indeed practically unmeasurable on a map printed on ordinary paper. The projection properties are graphically invisible, and are only retained in the mathematical formulas and computation methods appropriate to the projection.

Consider a small circle of angular radius θ on a sphere of radius R. Its diameter in its own plane is $2R\sin\theta$, but the distance across the circle along the surface of the sphere is $2R\theta$ (in radian). If the circle is mapped so that its circumference is of correct length at a certain scale, its diameter measured on the map will come out too small, to the same scale, in the ratio $(\sin\theta)/\theta$, that is approximately $1 - (1/6)\theta^2$…. Now suppose we require the map to have no scale variation greater than $1/2\,000$, then θ must be less than $\sqrt{6/2\,000}$, approximately $1/18$ radian, about $3°$. This indicates the limitation of extent that can be covered on a single projection system, given that a limit is set to the acceptable scale variation.

* * * * * * * * * * *

Maps are basically a flat or planar representation of part or all of the earth's surface. The basic problem of map making is that it is impossible to develop a surface with double curvature, such as a sphere or ellipsoid, onto a plane surface without distortion of some kind. Different map projections are designed to maintain some property of the ellipsoidal surface undistorted. The property held "true" or undistorted will depend on the purpose for which the map is designed. The different possibilities are:

(1) Area—The ratio between various areas on the ellipsoid and the projection is unchanged — **equal area projection.**

(2) Distance—Distance on the ellipsoid are projected without distortion—**equidistant projection.**

(3) Shape—Small shapes are correctly represented on the projection—**conformal projection.**

The attainment of any of the above properties is at the cost of distortions of the other properties. For example, shapes must be distorted to maintain correct representation of area in an equal area projection.

Projections also classified according to the type of projection surface used. The three commonly used projection surfaces are the **plane, the cone, and the cylinder.** The cone and the cylinder may be developed without distortion onto a plane if they are cut in a straight line along their length. The point on the line or lines at which these surfaces are tangential to or cut the ellipsoid are usually selected so that the area of interest will be near the center of the projection.

Historically, maps were generated geometrically by simply projecting surface of the ellipsoid onto the projection surface, where the point of projection governs many of the properties of the resulting map. In modern maps, the relationship between ellipsoidal coordinates and the resulting map coordinates is purely analytical.

For the surveyor, the most significant and useful map projections are **conformal projections**. In such a projection, the fact that small shapes are correctly represented means that the scale in all direction from a point is independent of azimuth and constant for a short distance. It follows that angular relationship at a point are correct, i.e., angles measured from or plotted on a conformal projection are the same as those that would be measured or set out on the earth's surface. The commonly used conformal projections are the Transverse Mercator and the Lambert Conformal Conic. These projections will be treated in the following sections. Map coordinates form a rectangular plane or grid coordinate system. Calculations in a grid coordinate system are much simpler than those on an ellipsoid surface, because simple Euclidian geometry may be used rather than the complex formulae required to compute between geodetic coordinates and ellipsoidal azimuth and distance.

The use of grid coordinates based on map projection is very convenient in office plotting of cadastral surveys and engineering surveys of limited extent. Parallels of latitude and meridians of longitude may be plotted on a projection to form a graticule which, because of the projection distortion, will not be rectangular.

It is therefore much more convenient for the positions of points to be plotted on the map using the rectangular grid coordinate system rather than by latitude and longitude. Cadastral surveys, engineering surveys, or any horizontal survey of limited extent which are started and closed on geodetic stations of a control network can be computed on the grid by relative simple computations employing plane trigonometry. Further, each grid position can be converted to a latitude and longitude position on the reference ellipsoid for integration into the control network using the mathematical relationships between the two systems. Grids have long been considered essential on military maps for ease in defining position.

It is convenient and mistakes are less likely to occur if all grid coordinates are positive numbers. For this purpose, a **false origin** is usually specified to the south and west of the limits of the projection. From this point, all eastings and northings within the limits of the projection will be positive.

7.2 Projection Mathematics

In the most general terms, a map projection may be described as a pair of formulas:
$$N = f_1(B, L), E = f_2(B, L) \tag{7.1}$$
for expressing rectangular coordinates, Northings and Eastings, as functions of latitude and longitude. In some literature on surveying and map projections, the coordinates are denoted x and y, and many surveyors have been accustomed to the x coordinate being in the north-south direction, but there can be confusion, even between different surveyors, and it is best to use the N and E notation, as is now done in the Ordnance Survey, for instance.

The basic mathematical operations involved in map projections are as follows:

(1) Conversion of ellipsoidal coordinates to grid coordinates $(B, L) \rightarrow (E, N)$.

(2) Conversion of grid coordinates to ellipsoidal coordinates $(E, N) \rightarrow (B, L)$.

(3) The reduction of measured azimuth, angles, directions, and distances to the corresponding quantities on the grid.

Formulae for these conversions and reductions on the each map projection (for example, Transverse Mercator and Lambert projections) are given in numerous references, principally the manuals for the particular projections.

To be able to use the formulae derived Gauss-Kruger projection as direct as possible, it is expedient to introduce a special parameter q, called the **isometric latitude**, on the ellipsoid. Let us first consider the length differential dS on the ellipsoid. We get

$$(dS)^2 = (MdB)^2 + (rdl)^2 = r^2\left[\frac{M^2}{r^2}(dB)^2 + (dl)^2\right] \tag{7.2}$$

By defining the differential of the new parameter as

$$dq = \frac{M}{r}dB \tag{7.3}$$

(7.2) may be rewritten as

$$(dS)^2 = r^2[(dq)^2 + (dl)^2] \tag{7.4}$$

The isometric latitude q is clearly defined only in terms of the geodetic latitude B and is obtained by integrating (7.3). We obtain

$$q = \int_0^B \frac{M}{r}dB = \int_0^B \frac{M}{N\cos B}dB = \ln\tan\left(\frac{\pi}{4} + \frac{B}{2}\right)\left(\frac{1-e\sin B}{1+e\sin B}\right)^{e/2} \tag{7.5}$$

A conformal mapping of an ellipsoid must clearly satisfy the following equations (cf. "Formulae used in the Transverse Mercator Projection"):

$$\frac{\partial N}{\partial q}\cdot\frac{\partial N}{\partial l} + \frac{\partial E}{\partial q}\cdot\frac{\partial E}{\partial l} = 0, \quad \left(\frac{\partial N}{\partial q}\right)^2 + \left(\frac{\partial E}{\partial q}\right)^2 = \left(\frac{\partial N}{\partial l}\right)^2 + \left(\frac{\partial E}{\partial l}\right)^2 \tag{7.6}$$

That is:

$$\frac{\partial N}{\partial q} = \frac{\partial E}{\partial l}, \quad \frac{\partial E}{\partial q} = -\frac{\partial N}{\partial l} \tag{7.7}$$

7.3 Conformal Map Projection

A map projection on which the shape of any small area of the surface mapped is preserved unchanged, and all angles around any point are correctly represented, that is, if two lines on the spheroid intersect at a certain angle, then the lines representing them on the projection will intersect at that same angle. Projections having this property are called **conformal projection**. Also called orthomorphic map projection.

This is a very useful property because it simplifies the practical use of such a projection.

7.3.1 Lambert Projection

The Lambert projection is a conformal conic projection with either one or two **standard parallels** of latitude. The geometric interpretation is that of a cone either tangential to or cutting the ellipsoid along one or two parallels of latitude. The two standard parallel case is analogous to the transverse mercator with the cylinder smaller than the ellipsoid. The **apex angle** (θ) of the cone is chosen so that the cone cuts the ellipsoid in the area to be mapped (Fig. 7.1).

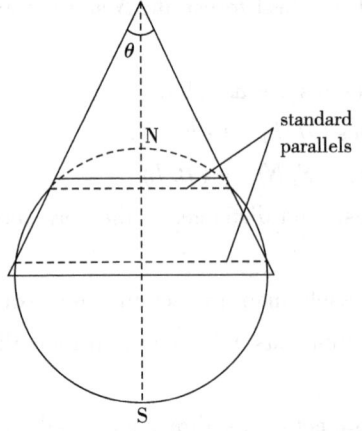

Fig. 7.1 Lambert conformal conic projection

The axis of the cone coincides with the earth's of rotation and the apex of the cone is either to the north or south of the earth depending on whether the area to be mapped is in the northern or southern hemisphere. When the cone is developed onto a plane, the relationship between the grid and the graticule is as shown in Fig. 7.2. Parallels project as circle arcs concave to the nearest pole, while meridians are straight lines converging toward the nearest pole. Meridians and parallels cross at right angles.

Standard parallels are usually chosen to divide the north-south extent of the projection area into three parts in the approximate ratio, 1/6, 2/3, 1/6 (Fig. 7.2).

Scale is constant in the east-west direction but varies in the north-south direction. It will be correct along the standard parallels, too large outside them and too small between them.

The variation of scale factor in the north-south direction for a typical Lambert projection is shown in Fig. 7.3.

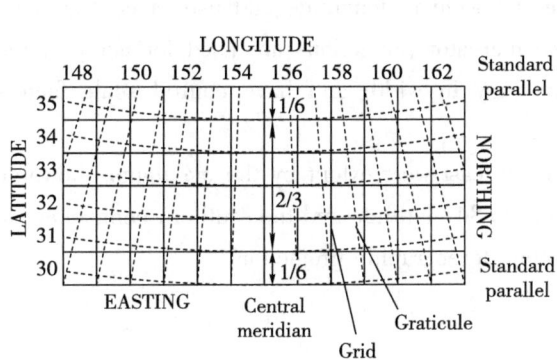

Fig. 7.2 Relationship between grid and graticule for a typical Lambert conformal conic projection

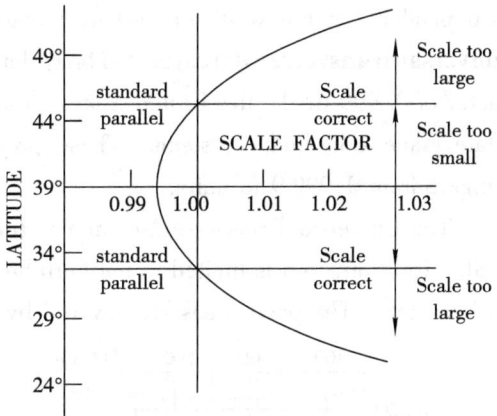

Fig. 7.3 Variation of scale with north coordinate on the Lambert conformal conic projection

7.3.2 Transverse Mercator Projection

The transverse Mercator is a conformal cylindrical projection and may be visualized as a cylinder wrapped around the earth and oriented so that its axis is in the plane of the equator. The cylinder is often slightly smaller than the earth in radius and intersects it along two ellipses equally spaced from and parallel to a **central meridian** of longitude (Fig. 7.4).

When the cylinder is developed onto a plane, the relationship between the grid and the graticule is shown in Fig. 7.5.

Meridians of longitude and parallels of latitude intersect at right angles. The central meridian is a straight line and the nearby meridians are nearly straight lines (slightly concave to the central meridian). The parallels are curved lines concave to the nearest pole.

The spacing between meridians of longitude, and hence the scale, increases away from the central

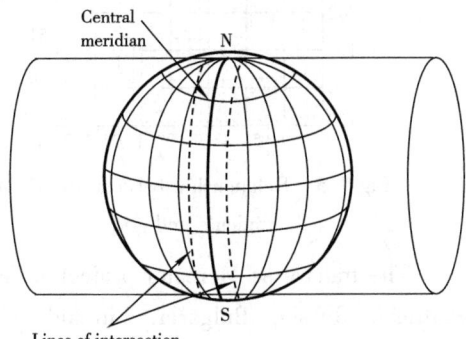

Fig. 7.4 The transverse Mercator projection

meridian. In order to preserve conformality, the scale in the north-south direction is distorted to equal the scale distortion in the east-west direction. The radius of the cylinder is chosen such that the scale distortion within the limits of the map sheet is kept to a minimum. The scale on the central meridian is too small as the cylinder is chosen to be smaller than the ellipsoid. Scale increases away from the meridian and is usually designed such that it is correct on two lines (ellipses of intersection) nearly parallel with and about two-thirds of the way between it and the edges of the projection. On the edges of the projection, the scale is too large. Fig. 7.6 shows the relationship between point scale factor and the east coordinate of a typical transverse mercator projection. Point scale factor is the ratio of an infinitesimal distance at a point on the grid to the corresponding distance on the ellipsoid. The value of scale factor on the central meridian, the central scale factor, is dependent on the width or east-west extent of the projection and the accuracy required. The **universal transverse Mercator (UTM)** has zones 6° wide in longitude and uses a central scale factor of 0.999 6. In the United States, transverse mercator projections are used for many of the State Plane Coordinate Systems. These projections vary in width and have central scale factors ranging from 0.999 9 to unity.

The Universal Transverse Mercator is limited in its east-west extent by the maximum allowable scale distortion, and is limited in northern hemisphere to 84° latitude and the southern hemisphere to 80° latitude. The polar areas are covered by the polar stereographic projection.

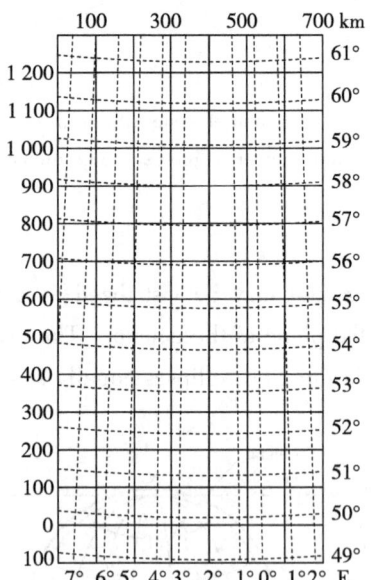

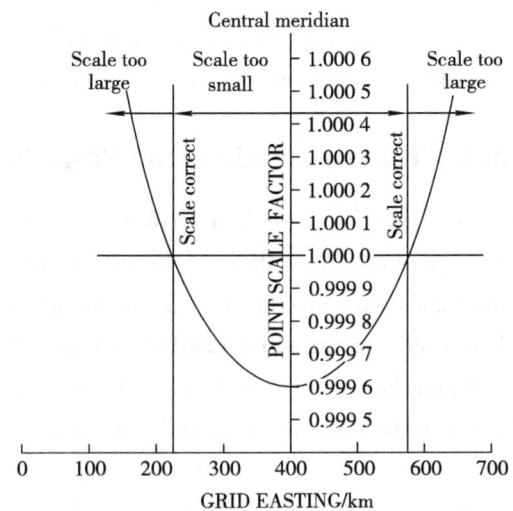

Fig. 7.5   Relationship between the UTM graticule and grid

Fig. 7.6   Variation of scale with east coordinate on the UTM

The transverse Mercator projection is used officially in Great Britain, Egypt, Sweden, Poland, Portugal, Russia, Bulgaria, Finland, Germany, Yugoslavia, Norway, British African Colonies, South Africa, Australia, U.S. Army Map Service, and in the plane coordinate systems of many States of the United States.

The transverse Mercator system is now more extensively used for geodetic computations than the Lambert conformal conic or any other projection.

7.3.3 Gauss-Krueger Projection

A map projection constructed in the same way as the Mercator projection but having the cylinder tangential to a meridian rather than to the equator. It is used mainly for small areas with a north-south orientation, and for all British Ordnance Survey maps. Rhumb lines on this projection are curved.

The transverse cylindrical orthomorphic system of projection, now usually called transverse Mercator, has also been called Lambert's cylindrical orthomorphic, and other names.

7.4 Transverse Mercator Projection

7.4.1 Formulae used in the Transverse Mercator Projection

A transverse projection is used for mapping an area of small extent in longitude; in the Universal Transverse Mercator it is 3° each side of a central meridian which is represented by a straight line on the projection. The usual structure for the Transverse Mercator has the central meridian true to scale, and the formulas for Northing and Easting linear coordinates are such that the representation is orthomorphic.

Let B be the spheroid latitude of a point and l its longitude reckoned from the central meridian. Then in practice l will be a small angle, not more than about 3°, or about 1/20 radian. The basic formulas for Northings and Eastings are expressed as series in powers of l (radian). Before plunging into concentrated mathematics, we can say a few things about the formulas:

(1) when l is zero, the formula for E must be zero, and the formula for N must be simply $N = S$, where S is the linear distance along the meridian, measured from the equator or from any specified point on the meridian (in the Ordnance Survey projection the structural zero point is 49° north latitude);

(2) since the projection is geometrically symmetrical with respect to the central meridian, the formula for N must be unchanged when l is changed to $-l$, and the formula for E must change sign without change of numerical value. Hence, we must be able to write:

$$N = S + Pl^2 + Ql^4 + Rl^6 + \ldots \qquad (7.8)$$
$$E = Al + \bar{B}l^3 + Cl^5 + \ldots$$

with only even powers of l in N and only odd powers in E. The coefficients $A, \bar{B}\ldots P, Q\ldots$ are functions of the latitude B, they have the dimension of length, and they incorporate the parameters-radius and eccentricity — that define the spheroid to which the survey system is referred.

What are the functions $A, \bar{B}\ldots P, Q\ldots$ that make the formulas for N and E and produce an orthomorphic transformation of the spheroid to a plane? We consider a simple example of differential geometry on the spheroid, and its representation on the plane. In Fig. 7.7, point A has latitude B

and longitude l. Let these change "infinitesimally" to $(B + dB)$ and $(l + dl)$: the dB means a linear shift of MdB along the meridian and the dl means a linear shift of $N\cos B dl$ along the parallel, to the point A'. In these formulas, M is the meridian radius of curvature $a(1 - e^2)(1 - e^2\sin^2 B)^{-3/2}$, and N is the other principle radius of curvature $a(1 - e^2\sin^2 B)^{-1/2}$, where a is the equatorial radius of the spheroid and e is the eccentricity.

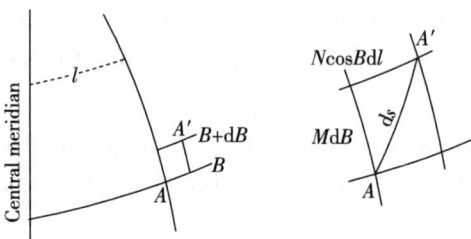

Fig. 7.7 The scale factor

Now, any projection can be expressed in general terms as two functions: $N = f_1(B, l)$, $E = f_2(B, l)$. If small changes dB and dl are made, the resulting changes of N and E are expressible by the differential formulas for two variables:

$$dN = \frac{\partial N}{\partial B}dB + \frac{\partial N}{\partial l}dl, \quad dE = \frac{\partial E}{\partial B}dB + \frac{\partial E}{\partial l}dl \tag{7.9}$$

The "infinitesimal" spheroid length is given by

$$(dS)^2 = M^2(dB)^2 + N^2\cos^2 B(dl)^2 \tag{7.10}$$

The corresponding infinitesimal length on the projection is ds given by

$$(ds)^2 = (dN)^2 + (dE)^2 \tag{7.11}$$

Thus the **"scale factor"** is ds/dS, and on substituting the differential formulas for dN and dE we get:

$$\left(\frac{ds}{dS}\right)^2 = \left\{\left[\left(\frac{\partial N}{\partial B}\right)^2 + \left(\frac{\partial E}{\partial B}\right)^2\right](dB)^2 + 2\left(\frac{\partial N}{\partial B}\cdot\frac{\partial N}{\partial l} + \frac{\partial E}{\partial B}\cdot\frac{\partial E}{\partial l}\right)(dB)(dl) + \left[\left(\frac{\partial N}{\partial l}\right)^2 + \left(\frac{\partial E}{\partial l}\right)^2\right](dl)^2\right\}/[M^2(dB)^2 + N^2\cos^2 B(dl)^2] \tag{7.12}$$

It is at this point that we go orthomorphic, by requiring that the above expression shall be independent of the particular values of dB and dl. That is, **the scale factor is to be independent of the direction; this is the definition of orthomorphism**.

Now, think of the two particular cases of $dB = 0$ and $dl = 0$; in the former case the square of the scale factor is simply

$$\left[\left(\frac{\partial N}{\partial l}\right)^2 + \left(\frac{\partial E}{\partial l}\right)^2\right]/N^2\cos^2 B \tag{7.13}$$

and in the latter case it is simply

$$\left[\left(\frac{\partial N}{\partial B}\right)^2 + \left(\frac{\partial E}{\partial B}\right)^2\right]/M^2 \tag{7.14}$$

Hence, for orthomorphism we certainly require these two fractions to be equal; but this is not enough. The expression for $(ds/dS)^2$ must be equal to the above fractions in general, that is when both dB and dl are non-zero. Clearly, this cannot be so unless the central term in the numerator of the general formula is identically zero. So we have the second condition for orthomorphism:

$$\frac{\partial N}{\partial B} \cdot \frac{\partial N}{\partial l} + \frac{\partial E}{\partial B} \cdot \frac{\partial E}{\partial l} = 0 \qquad (7.15)$$

The first condition may be written:

$$N^2 \cos^2 B \left[\left(\frac{\partial N}{\partial B} \right)^2 + \left(\frac{\partial E}{\partial B} \right)^2 \right] = M^2 \left[\left(\frac{\partial N}{\partial l} \right)^2 + \left(\frac{\partial E}{\partial l} \right)^2 \right] \qquad (7.16)$$

These equations are the well-known **Cauchy-Riemann conditions** for orthomorphic (or conformal) representation. They ensure that at any point in the projection the scale along infinitesimally short distances is the same in all directions.

It remains to apply these conditions to the series for N and E, and the desired formulas for A, $\bar{B}...P$, $Q...$ will be revealed. As mentioned above, these coefficients in the series are functions of B; it is also to be remembered that M is the meridian radius of curvature and by definition $MdB = dS\cos A$.

The four partial differential coefficients are:

$$\begin{aligned}
\frac{\partial N}{\partial l} &= 2Pl + 4Ql^3 + 6Rl^5 \ldots \\
\frac{\partial N}{\partial B} &= M + \frac{dP}{dB}l^2 + \frac{dQ}{dB}l^4 \ldots \\
\frac{\partial E}{\partial l} &= A + 3\bar{B}l^2 + 5Cl^4 \ldots \\
\frac{\partial E}{\partial B} &= \frac{dA}{dB}l + \frac{d\bar{B}}{dB}l^3 + \frac{dC}{dB}l^5 \ldots
\end{aligned} \qquad (7.17)$$

These are to be substituted into the Cauchy-Riemann conditions, taking care to keep terms in each power of l together. The algebra is somewhat heavy but quite straightforward, and the first few terms in the conditions will be found to be:

$$N^2 \cos^2 B \left\{ M^2 + \left[\left(\frac{dA}{dB} \right)^2 + 2M \frac{dP}{dB} \right] l^2 + \left[2 \frac{dA}{dB} \cdot \frac{d\bar{B}}{dB} + \left(\frac{dP}{dB} \right)^2 + 2M \frac{dQ}{dB} \right] l^4 \ldots \right\}$$
$$= M^2 [A^2 + (6A\bar{B} + 4P^2) l^2 + (10AC + 9\bar{B}^2 + 16PQ) l^4 \ldots] \qquad (7.18)$$

and

$$\left(2MP + A \frac{dA}{dB} \right) l + \left(4MQ + 2P \frac{dP}{dB} + 3\bar{B} \frac{dA}{dB} + A \frac{d\bar{B}}{dB} \right) l^3 + \ldots = 0 \qquad (7.19)$$

These are identities and must be satisfied separately in each power of l as well as by the terms not containing l. From the first identity we get $A = N\cos B$. Then from the second identity (term in l) we have $P = -(A/2M)(dA/dB)$. It is clear that the process of getting $\bar{B}, Q, C...$ will involve differentiating the succession of functions of B. For instance, $A = a\cos B(1 - e^2 \sin^2 B)^{-1/2}$, and the standard rules for differentiation show that its differential coefficient is $-M\sin B$, so we then find that $P = +\frac{1}{2}N\sin B \cos B$. Having then found that

$$\frac{dP}{dB} = \frac{1}{2} M \cos^2 B \left(\frac{N}{M} - \tan^2 B \right) \qquad (7.20)$$

We use the l^2 term in the first identity and get $\bar{B} = (1/6) N\cos^3 B [(N/M) - \tan^2 B]$. The ratio

N/M which turns up in some of the formulas is simply a neat way of writing $(1 - e^2\sin^2 B)/(1 - e^2)$ and in many cases it can be taken equal to 1 for practical computation.

After another bout of differentiation, Q comes from the coefficient of l^3; actually, the two inner terms in the bracket cancel and we find that

$$Q = \frac{1}{24}N\sin B\cos^3 B\left(4\frac{N^2}{M^2} + \frac{N}{M} - \tan^2 B\right) \tag{7.21}$$

At this point it is possible to start considering some simplification of formulas. The coefficient Q multiplies l^4: N is about 6.4×10^6 m, and the trigonometrical factors together have a maximum value of about 1.5 (at latitude 28°) so Q has a maximum of about 400 000 meters. With l limited to about $1/20$ radian, the term Ql^4 will rarely be as much as 2.5 m. Hence, for the reason mentioned above, we can take $Q = (1/24)N\sin B\cos^3 B(5 - \tan^2 B)$.

Of course, these simplifications were useful in the days of manual arithmetic and books of numerical tables. With electronic computers, the full formulas may well be used, and N/M must be kept in the l^3 term.

Finally, to get the coefficient C it is necessary to work out dQ/dB; anyone prepared to solg through all the calculus and algebra will find that C, in all its glory, is:

$$\frac{1}{120}N\cos^5 B\left[4\frac{N^3}{M^3} + \frac{N^2}{M^2} - \left(24\frac{N^3}{M^3} - 8\frac{N^2}{M^2} + 2\frac{N}{M}\right)\tan^2 B + \tan^4 B\right] \tag{7.22}$$

or in simpler garb:

$$\frac{1}{120}N\cos^5 B(5 - 18\tan^2 B + \tan^4 B) \tag{7.23}$$

giving Cl^5 a maximum of a few centimeters in practice. This is quite enough! All the rest of the terms in the series for N and E are too small to matter when the projection zone is 6° wide.

Thus, the formulas, in series taken as far as is necessary for working within a longitude zone of a few degrees, are:

$$N = S + \frac{N}{2}\sin B\cos Bl^2 + \frac{N}{24}\sin B\cos^3 B(5 - \tan^2 B)l^4\dots$$

$$E = N\cos Bl + \frac{N}{6}\cos^3 B\left(\frac{N}{M} - \tan^2 B\right)l^3 + \frac{N}{120}\cos^5 B(5 - 18\tan^2 B + \tan^4 B)l^5\dots \tag{7.24}$$

in which l is in radians, that is, $l(\text{radians}) = 1''/206\,264.8$. For calculating the meridian length S, a series in powers of e^2 must be used. The length from a small change dB is MdB, so the integral $\int_0^B MdB$ gives the meridian length from the equator to latitude B. (cf. the Length of the Meridian Arc)

7.4.2 Definitions and Relationship of UTM Projection Quantities

The following definitions and explanation of the relationship between quantities on the ellipsoid and the UTM projection is given as an example only.

A graphical representation of a UTM projection in the northern hemisphere is given in Fig. 7.8.

A straight line, P_1P_2, on the ellipsoid is generally a curved line on the projection. This curvature is practically always concave to the central meridian.

The **azimuth of the line** $P'_1 P'_2$ is shown as A_{12}. It is the clockwise angle measured at P'_1 between the ellipsoidal meridian at P'_1 and the point P'_2.

The **ellipsoidal distance,** S, of the line $P_1 P_2$, is the geodesic distance between P_1 and P_2 on the ellipsoid. The grid distance, s, of the line $P'_1 P'_2$ is the length of the projected arc between P'_1 and P'_2.

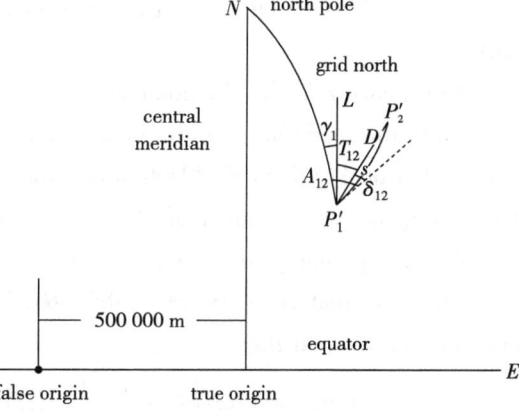

Fig. 7.8 The Universal Transverse Mercator Projection

The **plane distance,** D, of the line $P'_1 P'_2$ is the length of the chord joining the projected points, P'_1 and P'_2. The difference in length between the plane distance, D, and the grid distance, s, is nearly always negligible.

The **plane bearing,** T_{12}, from P'_1 to P'_2, is the clockwise angle at P'_1 between grid north and the chord joining the projected points, P'_1 and P'_2.

The **convergence of meridian,** γ_1, is the clockwise angle between true north and grid north at the point P'_1.

The **arc to chord correction,** δ_{12}, is the angle between the chord and the projected arc at the two ends of the line. The point scale factor, m, is the ratio of an infinitesimal distance at a point on the grid to the corresponding distance on the ellipsoid.

$$m = \frac{\mathrm{d}s}{\mathrm{d}S} \tag{7.25}$$

As the UTM is a conformal projection, the ratio is independent of the azimuth of the infinitesimal distance. The formulae for calculating the quantities defined above have not been given but are available in the various UTM manuals.

7.4.3 The Scale Factor and the Convergence of meridian

Two properties of a projection that vary from point to point are of practical importance, namely the scale factor and the convergence of meridian. The scale factor is $\mathrm{d}s/\mathrm{d}S$: formulas for $(\mathrm{d}s/\mathrm{d}S)^2$ as functions of the partial differential coefficients of N and E appear earlier in the "Formula used in the Transverse Mercator Projection" and either of them can now be put into terms of B and l by using the expressions found for $A, \bar{B}...P, Q....$ The result is

$$\frac{\mathrm{d}s}{\mathrm{d}S} = 1 + \frac{1}{2}\frac{N}{M}\cos^2 B l^2 + \ldots \tag{7.26}$$

Since $E = N\cos B l + \ldots$, the scale factor can be more usefully written as:

$$\frac{\mathrm{d}s}{\mathrm{d}S} = 1 + \frac{1}{2}\frac{E^2}{MN} + \ldots \tag{7.27}$$

At $E = 200$ km this amounts to $1.000\,491$. By doing more algebra, a term in E^4 can be added

—it is $E^4/24N^3M$ — but E must be more than 450 km before this amounts to as much as 1 part per million.

Grid bearing is the direction of a line referred to the coordinate system and true bearing is referred to the meridian; the difference between them is the convergence of meridian, and it is the angle LP'_1N on Fig. 7.8 (cf. "Definitions and Relationship of UTM Projection Quantities" Fig. 7.8). For a point on the meridian of P_1 there is no change of longitude, so the changes of the plane coordinates are simply $dE = (\partial E/\partial B)dB$ and $dN = (\partial N/\partial B)dB$. Therefore $\tan\gamma = -(dE/dN)$ when $dl = 0$, that is, $\tan\gamma = -(\partial E/\partial B)/(\partial N/\partial B)$. Using the series for the partial differential coefficients, we find that:

$$\tan\gamma = l\sin B\left(1 + \frac{1}{3}\frac{N}{M}l^2\ldots\right) \text{ and } \gamma = l\sin B + \frac{1}{3}l^3\sin B\cos^2 B + \ldots \quad (7.28)$$

For example, at $B = 50°$, $l = 3°$, the convergence of meridian is $2°17'53.28'' + 3.12'' = 2°17'56.4''$

****

In our country (China), the Gauss-Krueger projection has "true" scale on the central meridian, and the scale is larger elsewhere. In some countries, it is common practice to apply an overall reduction of scale so that the exaggeration at the boundaries of the mapped zone is reduced by about 50%. Then, if the reduction factor is $1/F$, the scale factor at easting E is $1 + (1/2)(E^2/MN) - 1/F$. For instance, if F is 2 000, the scale is true at about $E = 202$ km, and is 1/2 000 too large at about $E = 285$ km. The reduction factor may be put into effect by reducing the radius a of the spheroid before any linear quantities are calculated. Also, any measured lengths should be similarly reduced.

Vocabulary

conformal projection /kən'fɔːməl prə'dʒekʃən/ 正形投影
convergence of meridian /kən'vəːdʒəns ɔv mə'ridiːən/ *n.* 子午线收敛角
isometric latitude /ˌaisəu'metrik 'lætitjuːd/ 等量纬度
orthomorphic /ɔːθə'mɔːfik/ *adj.* 正形的
scale factor /skeil 'fæktə/ 长度比

8 Establishment of Geodetic Coordinate System

8.1 Datums

The individual countries, or in some cases, groups of countries, choose different reference ellipsoids. The factors that affect these selections are the size and shape of the ellipsoid as well as its position. Ellipsoids that have been defined with orientation and position as well as size and shape are referred to as **geodetic datums**. In the context of GPS positioning, these individual geodetic datums are often refered to as local datums, because in comparison with the satellite datums they only apply to a region or local area of the earth's surface.

There are two main satellite datums in general use: Precise Ephemeris Datum and Broadcast Ephemeris Datum.

8.1.1 Local Datums

The definition of a geodetic or local datum is usually quite arbitrary, and its selection is subject only to convenience. The size and shape of the ellipsoid must first be defined by selection of a semi-major axis length, a, and a flattening, f. The parameters chosen in the past have tended to depend on historic developments and international ellipsoids which have been agreed on from time to time. One control point, the origin, is chosen and the height of this point above the ellipsoid is defined. This definition may be arbitrary or it may be derived as a result of spirit leveling. The spirit leveling will be relative to mean sea-level (the geoid) and the use of the leveled height as the height above the ellipsoid implicitly defines the geoid to coincide with the ellipsoid at the origin. The position of the origin is also defined by geodetic latitude and longitude. One method of doing this is simply to adopt the observed astronomical latitude and longitude of the point. The **implication** of this method is that the ellipsoid beneath the point will be parallel to the horizontal plane at the point. In other words, there will be zero deviation of the vertical at the origin. It should be noted that either the geodetic azimuth or the geodetic longitude may be chosen, but not both, as Laplace's equation must be satisfied.

The definition is completed by choosing the orientation of the minor axis. This axis is always chosen to be parallel to the earth's axis of rotation, but because of polar motion, the direction of this rotation axis is time-dependent. The most convenient definition is the **Conventional International Origin** (CIO) pole.

A datum chosen by this method is not generally geocentric, as usually there has been no use of gravity observations to calculate the deviation of the vertical components (ξ, η) and the geoid-ellipsoid separation at the origin (N_0). Valuable information on orientation may be obtained from

Doppler observations.

8.1.2 Satellite Datums

Satellite datums are defined in a completely different way than local datums. Local datums deal with only part of the earth's surface and are unrelated or, at best, only loosely related to the earth's center of mass, the **geocenter**. The procedure outlined above for the definition of a local datum is completely different from that used in the definition of a satellite datum.

The satellite reference system is defined by the system in which the satellite ephemeris or orbit parameters are given. These orbit parameters are based on the adopted coordinates of a number of satellite tracking stations, an adopted geopotential model for the earth's gravity field and a set of constants.

These constants are:

(1) The Gravitational Constant times the earth's mass, GM.

(2) The rotation rate of the earth with respect to the instantaneous equinox, ω.

(3) The speed of light, c.

(4) Clock corrections and oscillator drift rates at the tracking stations used for ephemeris calculation.

The procedure for definition of a satellite datum is described (cf. Black, H. D., "The Transit System, 1977: performance, plans and potential", Meeting on Satellite Doppler Tracking and its Geodetic Applications at The Royal Society, Oct. 10-11, 1978, London, England.). The inclusion of a geopotential model as an integral part of the satellite datum means that the coordinate origin of the system is intended to be the earth's center of mass.

The concept a reference ellipsoid is unnecessary in the definition of a satellite datum, whereas it is a cornerstone in the local datum definition. The ellipsoid definition is not used in the orbit computation. However, an ellipsoid is usually associated with a satellite datum so that positions may be shown in geographic as well as in Cartesian coordinates. This ellipsoid is derived by a least squares fitting process. The leveled elevations of each of the tracking stations are subtracted from their Doppler derived radii to give points on the geoid. An ellipsoid is fitted to these points by least squares.

8.2 Geodetic Datum

A **geodetic datum,** normally referred to simply as datum, is defined by the location of the central point (**geodetic origin** or **datum origin**) with respect to the centre of the reference ellipsoid selected. This has been customarily achieved by astronomic latitude, longitude, and azimuth determinations at the geodetic origin; by defining the size and shape of the reference ellipsoid; and, preferably, by fulfilling the condition through the Laplace equation that the rotation axis of the ellipsoid be parallel to the mean rotation axis of the earth. If the deflection components and undulation of the geoid are absolute (i.e. not relative), then the datum is geocentric. Because the determination of these absolute values at the geodetic origin of a datum is difficult, each country commonly has had its own relative geodetic datum.

* * * * * * * * * * *

A **geodetic datum** is usually defined in terms of five parameters $a, f; x_0, y_0, z_0$. Here a and f denote semimajor axis and flattening of the reference ellipsoid, which is taken as an ellipsoid of revolution, and x_0, y_0, z_0 are the rectangular coordinates of the center of the reference ellipsoid with respect to the geocenter, the earth's center of mass. This definition presupposes an underlying basic system of rectangular coordinates XYZ, the Z-axis coinciding with the mean rotation axis of the earth, and the X-axis passing the zero meridian, which is the mean Greenwich meridian. The rotation axis of the ellipsoid is supposed to be parallel to the Z-axis.

* * * * * * * * * * *

A reference surface consisting of five quantities: the latitude and longitude of an geodetic, the azimuth of a line from this point, and two constants necessary to define the reference spheroid. It forms the basis for the computation of horizontal control surveys, in which the curvature of the earth is considered. Also called **horizontal control datum; horizontal datum; horizontal geodetic datum.**

* * * * * * * * * * *

The six topocentric parameters needed can be written as $B_0, L_0, h_0, \xi_0, \eta_0, A_0$; they play the following roles:

(1) B_0, L_0, h_0 are the three geodetic coordinates of the geodetic P_0, where h_0 is the sum of the orthometric height H_0 and the (relative) geoidal height N_0 referred to the datum to be positioned. The first parameters, B_0, L_0, specify a particular normal to the reference ellipsoid. Since the Local Astronomical system is fixed relative to the gravity field of the earth, once ξ_0 and η_0 are defined it follows that the normal to the reference ellipsoid (specified by B_0, L_0) is also fixed with respect to the earth. The choice of h_0 then fixes the depth of the ellipsoidal surface below or above the geodetic ($\varphi_0, \lambda_0, H_0$); at this stage, the ellipsoid is free only to rotate around the normal.

(2) A_0 is the geodetic azimuth, which should satisfy $\alpha_0 - A_0 = (\lambda_0 - L)\sin B$, where λ_0 and α_0 refer to the projection of P_0 onto the geoid. The choice of A_0 removes the last remaining degree of freedom of the reference ellipsoid.

(3) ξ_0, η_0, are the two (relative) geoidal deflection components at P_0 referred to the reference ellipsoid.

These six parameters are equivalent to the six geocentric parameters, three translations X_0, Y_0, Z_0 and three rotations $\varepsilon_X, \varepsilon_Y, \varepsilon_Z$. The advantage that the topocentric parameters have over the geocentric is that they have a direct relation to quantities measured on the surface of the earth.

* * * * * * * * * * *

It is important to realize that **maps are drawn and positions are defined with respect to a reference datum**. In the United States we use the North America Datum, in Japan the Tokyo Datum, in Europe the European Datum, etc. The GPS currently uses the World Geodetic System of 1984 (WGS-84). As a result, the same reference marker will have a different set of latitude and longitude coordinates in each reference datum. Apparent differences of 1/2 km occur in some locations.

The four parts of Fig. 8.1 help us visualize the concept of reference datums and how they relate to each other. We already indicated that the earth is an irregular shape due to density (gravity) variations, and Fig. 8.1(a) is an exaggerated model of an irregular "earth". The surface shown represents the geoid, which is defined as the location of mean sea level over the entire earth's surface.

In order to make reasonably accurate maps, a model of the earth's surface is needed. Fig. 8.1 (b) shows how such models have been designed to fit the earth over the area of local interest, which in the past never was larger than a continent. The model consists of a spheroid (ellipsoid) and one position called the datum at which latitude and longitude are defined. Such a model works well and allows accurate maps to be drawn in the vicinity of the datum.

Now that satellites are being used to measure the geoid (satellite geodesy), a different type of datum is needed. As illustrated by Fig. 8.1(c), a world spheroid may not fit the earth very well at any one location, but it is a "best fit" to the entire earth. In addition, there is not a single reference datum position because many satellite tracking stations are involved, and their positions are defined as part of the calculations which determine the earth's geopotential field (geoid). The WGS-84 shperoid is a "best fit" to the WGS-84 geoid.

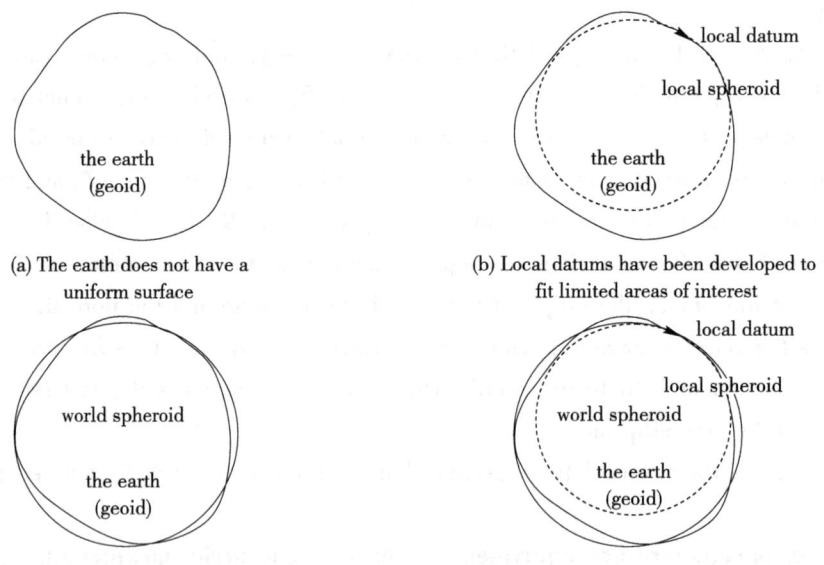

Fig. 8.1　Development and relationship of local and global reference datums

Fig. 8.1(d) makes it clear that there must be some method of relating a position in one datum to coordinates in another. For example, satellite position fixes taken in Tokyo harbor might show the ship to be well inland when plotted on a local chart. The reason is datum difference as illustrated by Fig. 8.1(d).

8.3 Coordinate Transformation and Datum Shifts

8.3.1 Mathematical Model

A geodetic datum is determined by the dimensions of the reference ellipsoid (semimajor a and flattening f) and its position with respect to the earth or the geoid. This relative position is usually given by the geoidal undulation N_1 and the components ξ_1 and η_1 of the deflection of the vertical at a geodetic origin P_1. Instead of ξ_1, η_1, N_1 we might as well use the geodetic coordinates B_1, L_1, h_1 of P_1 because

$$\begin{aligned} \xi_1 &= \varphi_1 - B_1 \\ \eta_1 &= (\lambda_1 - L_1)\cos B_1 \\ N_1 &= h_1 - H_1 \end{aligned} \qquad (8.1)$$

A superficially different but equivalent method is to use the rectangular coordinates x_0, y_0, z_0 of the center of the reference ellipsoid with respect to the center of the earth.

If we vary the geodetic datum—that is, the reference ellipsoid and its position—then the geodetic coordinates B, L, h and consequently the deflections of the vertical and the undulations of the geoid,

$$\begin{aligned} \xi &= \varphi - B \\ \eta &= (\lambda - L)\cos B \\ N &= h - H \end{aligned} \qquad (8.2)$$

will also change. Since there are three different ways of fixing the datum, we can formulate these changes in terms of the variation of

$$\xi_1, \eta_1, N_1, \text{ or } B_1, L_1, h_1, \text{ or } x_0, y_0, z_0$$

Mathematically, the problem is simply a transformation of coordinates, since every geodetic datum corresponds to a different system of geodetic coordinates B, L, h.

Suppose that the center of the reference ellipsoid does not coincide with the earth's center of gravity, but that the axis of the ellipsoid is parallel to the earth's axis of rotation. Assume a rectangular coordinate system XYZ whose origin is the earth's center of gravity (not the center of the ellipsoid), the axes being directed as usual. Let the coordinates of the center of the ellipsoid with respect to this system be x_0, y_0, z_0, as stated above. Then between (B, L, h) and (X, Y, Z) equations must obviously be modified so that they become

$$\begin{aligned} X &= x_0 + (N + h)\cos B \cos L \\ Y &= y_0 + (N + h)\cos B \sin L \\ Z &= z_0 + \left(\frac{b^2}{a^2}N + h\right)\sin B \end{aligned} \qquad (8.3)$$

These equations form the starting point for various important differential formulas of coordinate transformation. First we ask how the rectangular coordinates X, Y, Z change if we vary the geodetic coordinates B, L, h by small amounts $\delta B, \delta L, \delta h$ and if we also alter the geodetic datum, namely the reference ellipsoid (a, f) and its position (x_0, y_0, z_0), by $\delta a, \delta f$ and $\delta x_0, \delta y_0, \delta z_0$. Note that δx_0, $\delta y_0, \delta z_0$ correspond to a small translation (parallel displacement) of the ellipsoid, its axis remaining parallel to the axis of the earth. The solution of the problem is found by differentiating (8.3):

$$\delta X = \delta x_0 + \frac{\partial X}{\partial a}\delta a + \frac{\partial X}{\partial f}\delta f + \frac{\partial X}{\partial B}\delta B + \frac{\partial X}{\partial L}\delta L + \frac{\partial X}{\partial h}\delta h$$
$$\delta Y = \delta y_0 + \frac{\partial Y}{\partial a}\delta a + \frac{\partial Y}{\partial f}\delta f + \frac{\partial Y}{\partial B}\delta B + \frac{\partial Y}{\partial L}\delta L + \frac{\partial Y}{\partial h}\delta h \qquad (8.4)$$
$$\delta Z = \delta z_0 + \frac{\partial Z}{\partial a}\delta a + \frac{\partial Z}{\partial f}\delta f + \frac{\partial Z}{\partial B}\delta B + \frac{\partial Z}{\partial L}\delta L + \frac{\partial Z}{\partial h}\delta h$$

Since, according to Taylor's theorem, small changes can be treated as differentials.

In these differential formulas we shall be satisfied with an approximation. Since the flattening f is small, we may expand $N = a^2/b \sqrt{1 + e'^2\cos^2 B}$ as

$$N = \frac{a^2}{b}(1 + e'^2\cos^2 B)^{-1/2}$$
$$= \frac{a^2}{b}(1 - \frac{1}{2}e'^2\cos^2 B\ldots) \qquad (8.5)$$
$$= a(1 + f\ldots)(1 - f\cos^2 B\ldots)$$
$$= a(1 + f - f\cos^2 B\ldots)$$
$$N \doteq a(1 + f\sin^2 B) \qquad (8.6)$$

and

$$\frac{b^2}{a^2}N = (1 - 2f\ldots)a(1 + f\sin^2 B\ldots) \doteq a(1 - 2f + f\sin^2 B\ldots) \qquad (8.7)$$

since

$$b = a(1 - f), e'^2 = 2f\ldots \qquad (8.8)$$

Thus equations (8.3) are approximated by

$$X = x_0 + (a + af\sin^2 B + h)\cos B\cos L$$
$$Y = y_0 + (a + af\sin^2 B + h)\cos B\sin L \qquad (8.9)$$
$$Z = z_0 + (a - 2af + af\sin^2 B + h)\sin B$$

Now we can form the partial derivatives in (8.4), for instance

$$\frac{\partial X}{\partial a} = (1 + f\sin^2 B)\cos B\cos L \doteq \cos B\cos L \qquad (8.10)$$

since we may neglect the flattening in these coefficients. This amounts to using for the coefficients, and only for them, a spherical approximation is used. Similarly, all coefficients are easily obtained as partial derivatives, and equations (8.4) become

$$\delta X = \delta x_0 - a\sin B\cos L\delta B - a\cos B\sin L\delta L + \cos B\cos L(\delta h + \delta a + a\sin^2 B\delta f) \quad (8.11a)$$
$$\delta Y = \delta y_0 - a\sin B\sin L\delta B + a\cos B\cos L\delta L + \cos B\sin L(\delta h + \delta a + a\sin^2 B\delta f) \quad (8.11b)$$
$$\delta Z = \delta z_0 + a\cos B\delta B + \sin B(\delta h + \delta a + a\sin^2 B\delta f) - 2a\sin B\delta f \quad (8.11c)$$

These formulas give the changes in the rectangular coordinates X, Y, Z in terms of the variation in the position (x_0, y_0, z_0) and the dimensions (a, f) of the ellipsoid and in the geodetic coordinates B, L, h referred to it.

8.3.2 Transformation of the Geodetic Coordinates

Several important formulas for the transformation of coordinates may be derived from equations (8.11). First, let the position of P in space remain unchanged; that is, let

$$\delta X = \delta Y = \delta Z = 0$$

Determine the change of the geodetic coordinates B, L, h if the dimensions of the reference ellipsoid and its position are varied.

The problem is thus to solve equations (8.11) for $\delta B, \delta L, \delta h$, the left-hand sides being set equal to zero. To get δB, multiply (8.11a) by $-\sin B\cos L$, (8.11b) by $-\sin B\sin L$, and (8.11c) by $\cos B$, and add all equations obtained in this way. For δL the factors are $-\sin L, \cos L$ and 0; for δh they are $\cos B\cos L, \cos B\sin L$ and $\sin B$. The result is

$$a\delta B = \sin B\cos L \delta x_0 + \sin B\sin L \delta y_0 - \cos B \delta z_0 + 2a\sin B\cos B \delta f$$
$$a\cos B \delta L = \sin L \delta x_0 - \cos L \delta y_0 \qquad (8.12)$$
$$\delta h = -\cos B\cos L \delta x_0 - \cos B\sin L \delta y_0 - \sin B \delta z_0 - \delta a + a\sin^2 B \delta f$$

We have seen that the translation of the ellipsoid may be also given in terms of the changes in the geodetic coordinates $\delta B, \delta L, \delta h$ of a geodetic origin, instead of $\delta x_0, \delta y_0, \delta z_0$. The problem is then to determine the variations $\delta B, \delta L, \delta h$ at the other points.

First we express the parallel displacement (x_0, y_0, z_0) of the ellipsoid in terms of the given $\delta B, \delta L, \delta h$. In equations (8.11), set $\delta X = \delta Y = \delta Z = 0$ (again because the position of the points in space remains unchanged) and $B = B_1, L = L_1, h = h_1$. Then we get

$$\delta x_0 = a\sin B_1 \cos L_1 \delta B_1 + a\cos B_1 \sin L_1 \delta L_1 - \cos B_1 \cos L_1(\delta h_1 + \delta a + a\sin^2 B_1 \delta f)$$
$$\delta y_0 = a\sin B_1 \sin L_1 \delta B_1 - a\cos B_1 \cos L_1 \delta L_1 - \cos B_1 \sin L_1(\delta h_1 + \delta a + a\sin^2 B_1 \delta f)$$
$$\delta z_0 = -a\cos B_1 \delta B_1 - \sin B_1(\delta h_1 + \delta a + a\sin^2 B_1 \delta f) + 2a\sin B_1 \delta f$$

$$(8.13)$$

These expressions for the shift components x_0, y_0, z_0 are inserted into equations (8.12), so that we finally obtain:

$$\delta B = (\cos B_1 \cos B + \sin B_1 \sin B\cos\Delta L)\delta B_1 - \sin B\sin\Delta L\cos B_1 \delta L_1 +$$
$$(\sin B_1 \cos B - \cos B_1 \sin B\cos\Delta L)\left(\frac{\delta h_1}{a} + \frac{\delta a}{a} + \sin^2 B_1 \delta f\right) +$$
$$2\cos B(\sin B - \sin B_1)\delta f$$
$$\cos B\delta L = \sin B_1 \sin\Delta L \delta B_1 + \cos\Delta L\cos B_1 \delta L_1 - \qquad (8.14)$$
$$\cos B_1 \sin\Delta L\left(\frac{\delta h_1}{a} + \frac{\delta a}{a} + \sin^2 B_1 \delta f\right)$$
$$\frac{\delta h}{a} = (\cos B_1 \sin B - \sin B_1 \cos B\cos\Delta L)\delta B_1 + \cos B\sin\Delta L\cos B_1 \delta L_1 +$$
$$(\sin B_1 \sin B + \cos B_1 \cos B\cos\Delta L)\left(\frac{\delta h_1}{a} + \frac{\delta a}{a} + \sin^2 B_1 \delta f\right) -$$

$$\frac{\delta a}{a} + (\sin^2 B - 2\sin B_1 \sin B)\delta f$$

where

$$\Delta L = L - L_1$$

These formulas express the variations δB, δL, δh at an arbitrary point in terms of the variations δB_1, δL_1, δh_1 at a given point and the changes δa and δf of the parameters of the reference ellipsoid. They thus relate two different systems of geodetic coordinates, provided these systems are so close to each other that their differences may be considered as linear. Mathematically, equations (8.14) are infinitesimal coordinate transformations; to the geodesist, they give the effect of a change in the geodetic datum. They are equivalent to equations (8.12). Both (8.12) and (8.14) are infinitesimal transformations of geodetic coordinates; they differ only in the parameters used for determining the coordinate system, the geodetic datum; in (8.12) the coordinate system is defined by (a, f, x_0, y_0, z_0) and in (8.14) by (a, f, B_1, L_1, h_1).

8.3.3 Transformation of ξ, η, N

Usually, equations (8.14) are expressed in terms of the variations of the deflection components ξ and η and of the geoidal undulation N. Since the natural coordinates φ, λ, H are not affected by a datum shift and remain unchanged, we get from (8.2)

$$\begin{aligned}\delta B &= -\delta\xi \\ \delta L\cos B &= -\delta\eta \\ \delta h &= \delta N\end{aligned} \tag{8.15}$$

so that equations (8.14) assume the form

$$\begin{aligned}\delta\xi &= (\cos B_1 \cos B + \sin B_1 \sin B \cos\Delta L)\delta\xi_1 - \sin B\sin\Delta L\delta\eta_1 - \\ &\quad (\sin B_1 \cos B - \cos B_1 \sin B \cos\Delta L)\left(\frac{\delta N_1}{a} + \frac{\delta a}{a} + \sin^2 B_1 \delta f\right) - \\ &\quad 2\cos B(\sin B - \sin B_1)\delta f \\ \delta\eta &= \sin B_1 \sin\Delta L\delta\xi_1 + \cos\Delta L\delta\eta_1 + \\ &\quad \cos B_1 \sin\Delta L\left(\frac{\delta N_1}{a} + \frac{\delta a}{a} + \sin^2 B_1 \delta f\right) \\ \frac{\delta N}{a} &= -(\cos B_1 \sin B - \sin B_1 \cos B \cos\Delta L)\delta\xi_1 - \cos B\sin\Delta L\delta\eta_1 + \\ &\quad (\sin B_1 \sin B + \cos B_1 \cos B \cos\Delta L)\left(\frac{\delta N_1}{a} + \frac{\delta a}{a} + \sin^2 B_1 \delta f\right) - \\ &\quad \frac{\delta a}{a} + (\sin^2 B - 2\sin B_1 \sin B)\delta f\end{aligned} \tag{8.16}$$

These formulas for the effect of a shift of the geodetic datum are **among the most important equations of geodesy**. They were derived by several scientists, among them Vening Meinesz (1950, 1953), by whose name they are usually known. Superficially similar formulas due to Helmert were in use earlier, but they are based on completely different geometrical principles, and are not suited for the purposes of modern geodesy.

8.3.4 Applications

As an illustration, we shall apply these formulas to the most important practical case, the absolute orientation of a local geodetic system, or its conversion to a world geodetic system. Let us assume that a triangulation or trilateration net has been computed on a local geodetic datum ($a', f', \xi_1', \eta_1', N_1'$). The quantities referred to this system will be indicated by a prime. Thus ξ_1', η_1', N_1' belong to the fundamental point P_1; they may be assumed to be zero or to have any other values.

Suppose now that at the geodetic origin the absolute geoidal height N_1 and the absolute deflection components ξ_1 and η_1 are known. The absolute values N, ξ, η will in general refer to a different ellipsoid (a, f), whose center is at the center of gravity of the earth. The quantities a, f, ξ_1, η_1, N_1 determine this "world geodetic system" completely.

It is now very easy to transform the local system ($a', f', \xi_1', \eta_1', N_1'$) to the world system. Set

$$\begin{aligned} \delta\xi_1 &= \xi_1 - \xi_1' \\ \delta\eta_1 &= \eta_1 - \eta_1' \\ \delta N_1 &= N_1 - N_1' \\ \delta a &= a - a' \\ \delta f &= f - f' \end{aligned} \qquad (8.17)$$

and compute, for all points of the local system, the changes $\delta\xi, \delta\eta, \delta N$ by equations (8.16). Then the ξ, η, N in the world system are given by

$$\begin{aligned} \xi &= \xi' + \delta\xi \\ \eta &= \eta' + \delta\eta \\ N &= N' + \delta N \end{aligned} \qquad (8.18)$$

The geodetic coordinates in the world geodetic system are obtained from

$$\begin{aligned} B &= B' - \delta\xi \\ L &= L' - \delta\eta \sec B \\ h &= h' + \delta N \end{aligned} \qquad (8.19)$$

and the geocentric rectangular coordinates X, Y, Z can be computed by the relationship (B, L, h) and (X, Y, Z) equations. A related problem is that of determining the coordinates x_0', y_0', z_0' of the center of the original reference ellipsoid defining the local datum ($a', f', \xi_1', \eta_1', N_1'$). Since the new datum ($a, f, \xi_1, \eta_1, N_1$), the world datum, is in an absolute position, we have

$$x_0 = y_0 = z_0 = 0$$

so that

$$\begin{aligned} \delta x_0 &= x_0 - x_0' = -x_0' \\ \delta y_0 &= y_0 - y_0' = -y_0' \\ \delta z_0 &= z_0 - z_0' = -z_0' \end{aligned} \qquad (8.20)$$

and

$$x_0' = -\delta x_0, \quad y_0' = -\delta y_0, \quad z_0' = -\delta z_0 \qquad (8.21)$$

where $\delta x_0, \delta y_0, \delta z_0$ are computed from (8.13). This solves our problem.

8.4 Orientation of Astrogeodetic Systems, Best Fitting Ellipsoids

The fit of an astrogeodetic system to the geoid may be improved by determining the deflections of the vertical and geoid undulations for a large number of network points, instead of for just one origin point.

In the three-dimensional projective method, three spatial displacements of the ellipsoid with respect to the geoid and a variation in the ellipsoidal parameters are permitted; the parallelism of the geodetic and global systems should be maintained. The corresponding changes $\delta\xi_1$, $\delta\eta_1$, δN_1, δa, δf in the parameters of the geodetic datum create changes in the ellipsoidal coordinates and, also in the deflection of the vertical $\delta\xi$, $\delta\eta$ and in the geoid undulation δN of all network points. $\delta\xi$, $\delta\eta$ and δN are given by the projective equations of the deflection of the vertical of F. A. Vening-Meinesz (1950), (spherical approximation, cf. "Coordinate Transformation and Datum Shifts" (8.16)).

The deflections ξ', η', or geoid undulations N' may be available for a number of points of a particular geodetic datum. We set

$$\xi = \xi' + \delta\xi, \quad \eta = \eta' + \delta\eta, \quad N = N' + \delta N \qquad (8.22)$$

where $\delta\xi$, $\delta\eta$, and δN are given by (8.16) (cf. "Coordinate Transformation and Datum Shifts").

The condition for a minimum

$$\sum (\xi^2 + \eta^2) = \min \qquad (8.23)$$

or, the equally valid condition

$$\sum N^2 = \min \qquad (8.24)$$

then furnishes the correction to the geodetic datum. Subsequently, (8.16) reveals the changes in ξ, η, N at every point.

The two-dimensional translational method, developed by Helmert, uses quantities which are reduced to the ellipsoid. A network displacement ($\delta\xi_1$, $\delta\eta_1$), an associated rotation of the network due to the Laplace equation, a scale change, and an ellipsoid transition (δa, δf) are permitted in this method. The relationship between the changes $\delta\xi$, $\delta\eta$ in the network points and the corrections to the datum is given by the translational equations of the deflection of the vertical corresponding to (8.16). An adjustment of the deflections of the vertical using $\sum (\xi^2 + \eta^2) = \min$. yields the changes in the datum.

Employing areally distributed measurements (area method), J. F. Hayford (1909) and T. N. Krassowski (1942), among others, determined ellipsoids from adjustments of the deflections of the vertical. In this respect, by assuming the isostatic compensation theory of Pratt, Hayford (1909) corrected the observed deflections in the U. S. A. for the influence of the topographic masses; he thus obtained ellipsoidal parameters that are unperturbed by any irregular masses. Krassowski used astrogeodetic material from the U. S. S. R., U. S. A., and Europe in computing a triaxial ellipsoid. The parameters $a = 6\,378\,245$ m, $f = 1/298.3$ determine a biaxial ellipsoidal which is the basis for geodetic surveys in East European countries.

Computations for a geodetic datum from astrogeodetic deflections and geoid undulations can only be carried out on the continents. They provide ellipsoids which best fit the dimensions and position of the geoid in their respective regions (**best fitting ellipsoids**); the mean earth ellipsoid can not be determined in this way. A geocentric position of these systems is as well not obtainable. Nevertheless, the deflections of the vertical or geoid undulations remain small in an optimally fitted region, so that there are only small reductions of the distance and horizontal directions to the ellipsoid.

Since geocentric coordinates for a great number of points are available today, along with good values for the mean earth ellipsoid, the methods for determining astrogeodetic datums have lost their significance. The system of equations (8.16) serves in the transformation to a given global datum.

8.5 Three-dimensional Transformations

Three-dimensional transformations are more suitable for use with Doppler positioning for a number of reasons. They are typically global in concept, they enable solutions for height as well as horizontal position, and they are mathematically rigorous. The complete three-dimensional transformation involves seven parameters that relate Cartesian coordinates in the two systems. There are **three translation parameters** to relate the origins of the two systems ($\Delta X, \Delta Y, \Delta Z$), **three rotation parameters**, one around each of the coordinate axis ($\omega_X, \omega_Y, \omega_Z$), to relate the orientation of the two systems, and **one scale parameter** (Δ) to account for any difference in scale between the two systems.

There are two widely used models for three-dimensional transformations, the Bursa-Wolf and the Molodensky-Badekas models.

The formulation of the **Bursa-Wolf model** is as follows:

$$\begin{pmatrix} X_2 \\ Y_2 \\ Z_2 \end{pmatrix} = \begin{pmatrix} \Delta X \\ \Delta Y \\ \Delta Z \end{pmatrix} + (1 + \Delta) \begin{pmatrix} 1 & -\omega_Z & \omega_Y \\ \omega_Z & 1 & -\omega_X \\ -\omega_Y & \omega_X & 1 \end{pmatrix} \begin{pmatrix} X_1 \\ Y_1 \\ Z_1 \end{pmatrix} \qquad (8.25)$$

where X_2, Y_2, Z_2 are coordinates in the satellite datum, and X_1, Y_1, Z_1 are coordinates in the local datum.

This model is well-suited for the comparison of and transformation between two satellite datums. It is not very well suited for transformation from a satellite datum to a local datum as the regional (rather than global) nature of the stations used in the computation of parameters gives rise to high correlation between the parameters. In other words, over a limited area it is hard to distinguish between offsets due to translation ($\Delta X, \Delta Y, \Delta Z$) and those due to rotation ($\omega_X, \omega_Y, \omega_Z$). As a result, large errors can occur which offset each other in the local area but which rapidly build up large datum shift discrepancies outside the local area.

The **Molodensky-Badekas model** overcomes the correlation problem by relating the scale and rotation parameters to some fundamental point, M, and working with differences from this point.

$$\begin{pmatrix} X_2 \\ Y_2 \\ Z_2 \end{pmatrix} = \begin{pmatrix} \Delta X \\ \Delta Y \\ \Delta Z \end{pmatrix} + \begin{pmatrix} X_M \\ Y_M \\ Z_M \end{pmatrix} + \begin{pmatrix} 1+\Delta & -\omega_Z & \omega_Y \\ \omega_Z & 1+\Delta & -\omega_X \\ -\omega_Y & \omega_X & 1+\Delta \end{pmatrix} \begin{pmatrix} X_1 - X_M \\ Y_1 - Y_M \\ Z_1 - Z_M \end{pmatrix} \quad (8.26)$$

Rearranging:

$$\begin{pmatrix} X_2 \\ Y_2 \\ Z_2 \end{pmatrix} = \begin{pmatrix} \Delta X \\ \Delta Y \\ \Delta Z \end{pmatrix} + \begin{pmatrix} X_1 \\ Y_1 \\ Z_1 \end{pmatrix} + \begin{pmatrix} \Delta & -\omega_Z & \omega_Y \\ \omega_Z & \Delta & -\omega_X \\ -\omega_Y & \omega_X & \Delta \end{pmatrix} \begin{pmatrix} X_1 - X_M \\ Y_1 - Y_M \\ Z_1 - Z_M \end{pmatrix} \quad (8.27)$$

The Defense Mapping Agency recommends that the fundamental point (X_M, Y_M, Z_M) is the average local position of the common points.

$$\begin{aligned} X_M &= \sum_{i=1}^{n} \frac{X_i}{n} \\ Y_M &= \sum_{i=1}^{n} \frac{Y_i}{n} \\ Z_M &= \sum_{i=1}^{n} \frac{Z_i}{n} \end{aligned} \quad (8.28)$$

The correlation between parameters in the Bursa-Wolf model and the decrease in correlation using the Molodensky-Badekas model is illustrated (cf. Boucher, C., "Investigation on Geodetic Applications of Satellite Doppler Observations for Control Networks", Proceedings of the Second International Geodetic Symposium on Satellite Doppler Positioning, Austin, Texas, January 1979.)

The seven parameters for either the Bursa-Wolf or the Molodensky-Badekas models can be solved for in a least squares adjustment. The adjustment uses points with coordinates in both the satellite and local datum along with their estimated variances and derives least squares estimates of the seven transformation parameters to fit the differences between the two sets of coordinates. The reliability of these derived parameters is usually expressed in terms of standard deviation or variance.

It is important that the parameters be as uncorrelated as possible when looking at the magnitudes and standard deviations of the different parameters. Where the standard deviations of particular parameters are equal to or larger than the parameters themselves, there is strong justification for omitting them. For instance, a rotation of 0.07 arc seconds with a standard deviation of ±0.19 arc seconds cannot be considered significant. If this parameter is neglected in a Bursa-Wolf transformation, the values of the remaining parameters will change because of the correlation. In the Molodensky-Badekas model, the remaining parameters would be unchanged.

In practice, a situation often occurs in which the number of points common to the two datums is insufficient for the accurate determination of all seven parameters. For example, if there are three common points, all seven parameters may be determined, but with only two degrees of freedom in the solution. In a case like this, the most logical solution is simply to determine the average translations (X, Y, Z) for the three points and be content with a three-parameter transformation.

These translations may also be determined by least squares using the simple model:

$$\begin{pmatrix} X_2 \\ Y_2 \\ Z_2 \end{pmatrix} = \begin{pmatrix} X \\ Y \\ Z \end{pmatrix} + \begin{pmatrix} X_1 \\ Y_1 \\ Z_1 \end{pmatrix} \qquad (8.29)$$

This model corresponds to the Bursa-Wolf or the Molodensky-Badekas models with the rotations and the scale bias constrained to be zero. This will usually be satisfactory as long as the area for which the transformations are to be used is small. As the area increases, the effect of the unknown rotations or scale change will become significant and lead to distortions.

If a three-parameter transformation is to be used, it may be desired to express it in terms of (ΔB, ΔL, Δh) rather than (ΔX, ΔY, ΔZ) so that geodetic coordinates may be directly transformed from one system to another. The Standard or Abridged Molodensky formula enables (ΔB, ΔL, Δh) to be calculated given (ΔX, ΔY, ΔZ), the approximate position (B, L, h) and the dimensions of the two ellipsoids (DMA, "Satellite Records Manual-Doppler Geodetic Point Positioning", DMA TM T-3-52320, Washington, 1976.).

As the Cartesian coordinates (X, Y, Z) are the system of coordinates used in GPS positioning, it is usually more convenient to simply transform the Cartesian coordinates of a point using (ΔX, ΔY, ΔZ) and then use the formulae given in section (cf. "Geodetic and Cartesian Coordinates") to calculate the corresponding geodetic coordinates (B, L, h). Care should be taken when using a three parameter transformation not to extrapolate far outside the area of the common points used to derive the transformation. Ideally, the transformation should only be used for points within the area defined by the common points (Fig. 8.2).

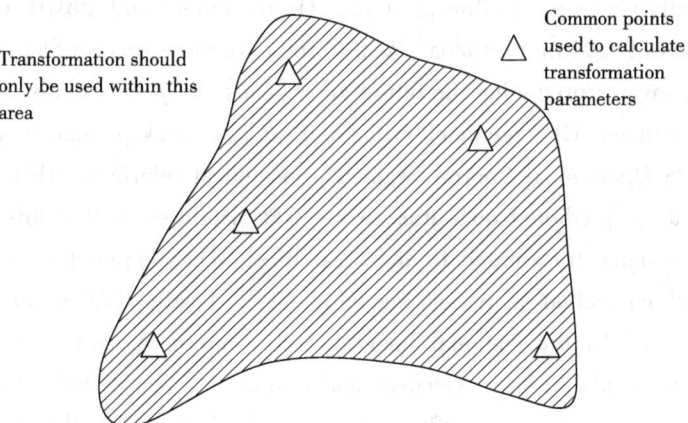

Fig. 8.2   A three-parameter transformation applies within the area of common points

The three-parameter transformation implicitly assumes that there are no rotations and no scale change between the two systems. If rotations or a scale change do exist, their effects are being accommodated by the three translation parameters calculated by averaging the ΔX, ΔY, ΔZ differences at the common points. If the residuals from this average three-parameter transformation are reasonable, it can be assumed that any rotations or a scale change are being adequately accounted for within the area of the common points. If we move outside the area of the common points, the transformation may no longer be valid, as the rotations and scale change may well become significant. To illustrate this, consider the simplest case where there is only one common point. The ΔX, ΔY, ΔZ derived from this point will give an error-free transformation for this point even though large rotations and a scale change may exist between the two systems. Away from this point, the rotations and scale change becomes significant, and the transformation is no longer valid. If we have more than one common point, we can inspect the residuals from the average ΔX, ΔY, ΔZ and determine whether or not the transformation is accurate enough for the intended purpose within the area contained by the common points.

8.6 World Geodetic System 1984

The Department of Defense (DoD) developed the WGS-84 reference system to support global activities involving mapping, charting, positioning, and navigation. More specifically, DoD introduced WGS-84 to express satellite orbits; that is, satellite positions as a function of time. Accordingly, WGS-84 is widely used for "absolute" positioning activities whereby people assume that satellite orbits are sufficiently accurate to serve as the sole source of control for positioning points of interest. In particular, absolute positioning does not rely on using positional coordinates for pre-existing terrestrial points for control. The general user, however, never needs to know the positions of these tracking stations.

DoD provides both "predicted" and "postfit" orbits in the WGS-84 reference system. As implied by the name, **predicted orbits** are calculated ahead of time by applying physical principles to extrapolate currently observed satellite positions. On the other hand, **postfit orbits** are calculated from previously observed satellite positions. Postfit orbits are more precise than predicted orbits both because they do not involve predicting the future and because they are usually derived using a larger number of tracking stations. GPS predicted orbits and satellite clock parameters are generated by the Air Force at the GPS Operational Control Segment, located at Schriever AFB, Colorado. The Air Force then uploads these predicted quantities to the GPS satellites so that this information may be included in the radio signal transmitted by these satellites. These predicted orbits support all real-time positioning and navigation activities involving GPS. Postfit GPS orbits and satellite clock parameters are generated by the National Imagery and Mapping Agency (NIMA), who currently makes this information available on its Geodesy and Geophysics World Wide Web pages. A number of other organizations also generate postfit GPS orbits which they usually express in a particular realization of the International Terrestrial Reference System (ITRS).

The original WGS-84 realization essentially agrees with NAD 83 (1986). Subsequent WGS-84 realizations, however, approximate certain ITRS realizations. Because GPS satellites broadcast the predicted WGS-84 orbits, people who use this broadcast information for positioning points automatically obtain coordinates that are consistent with WGS-84. Hence, the popularity of using GPS for real-time positioning has promoted greater use of WGS-84. Despite its popularity, people generally do not use WGS-84 for high-precision positioning activities, because such activities require the use of highly accurate positions on pre-existing terrestrial points for control. For example, various differential GPS techniques use known positions for one or more pre-existing terrestrial points to remove certain systematic errors in computing highly precise positions for new points. Consequently, before WGS-84 can support high-precision positioning activities, a rather extensive network of accurately positioned WGS-84 terrestrial control points would have to be established.

DoD established the original WGS-84 reference frame in 1987 using Doppler observations from the Navy Navigation Satellite System (NNSS) or TRANSIT. The WGS-84 frames have evolved significantly since the mid-1980s. In 1994, DoD introduced a realization of WGS-84 that is based completely on GPS observations, instead of Doppler observations. This new realization is officially known as WGS-84 (G730) where the letter G stands for "GPS" and "730" denotes the GPS week number (starting at 0h UTC, 2 January 1994) when NIMA started expressing their derived GPS orbits in this frame. The second WGS-84 realization, called WGS-84 (G873), is also based completely on GPS observations. Again, the letter G reflects this fact, and "873" refers to the GPS week number starting at 0h UTC, 29 September 1996. The latest realization of the WGS-84, designated as WGS-84 (G1150), is updated by NGA (National Geospatial-Intelligence Agency). The GPS Operational Control Segment implemented it on 20 January 2002. The implementation of improved methodologies has increased the accuracy of the coordinates of the U. S. Air Force (USAF) and NGA Global Positioning System (GPS) monitor stations.

The origin, orientation, and scale of WGS-84 (G873) are determined relative to adopted positional coordinates for 15 GPS tracking stations: five of them are maintained by the Air Force and ten by NIMA. NIMA chose their sites to complement the somewhat equatorial distribution of the Air Force sites and to optimize multiple station visibility from each GPS satellite.

People may anticipate further improvements of WGS-84 in the future, as new GPS tracking sites may be added or existing antennas may be relocated or replaced. NGA is dedicated to take appropriate measures to guarantee the highest possible degree of quality and to perpetuate the accuracy of WGS-84. As mentioned earlier, however, most regions lack a network of accessible reference points that might serve as control points from which highly accurate WGS-84 coordinates may be propagated using an appropriate static differential GPS technique involving carrier phase observables. More information about WGS-84 may be obtained via the Internet by accessing: http://164.214.2.59/GandG/tr8350_2.html.

8.7 ITRS and Its Realization

8.7.1 The Evolution of ITRS

In the late 1980s, the International Earth Rotation Service (IERS) introduced ITRS to support those scientific activities that require highly accurate positional coordinates; for example, monitoring crustal movement and the motion of earth's rotational axis. The initial ITRS realization was called the International Terrestrial Reference Frame of 1988 (ITRF88). Accordingly, IERS published positions and velocities for a worldwide network of several hundred stations. The IERS, with the help of several cooperating institutions, derived these positions and velocities using various highly precise geodetic techniques including GPS, VLBI, SLR, LLR, and DORIS (Doppler orbitography and radiopositioning integrated by satellite). Every year or so since introducing ITRF88, the IERS has developed a new ITRS realization—ITRF89, ITRF90, ..., ITRF2005—whereby they have published revised positions and velocities for previously existing sites, as well as new positions and velocities for those sites that had been established since after earlier realizations had been developed. Each new realization not only incorporated at least an additional year of data, but also the most current understanding of earth's dynamic behavior. The ITRF96 frame is defined by the positions and velocities of 508 stations dispersed among 290 globally distributed sites. Recall that a particular site may involve one or more co-located instruments employing various space-related techniques (e.g., GPS, VLBI, SLR, LLR, and DORIS). The accuracy and rigour of ITRS has proven contagious, and its popularity is steadily growing among those who engage in positioning activities.

Furthermore, ITRS is the first major international reference system to directly address plate tectonics and other forms of crustal movement by publishing velocities as well as positions for its control points. To appreciate the need for velocities, consider the theory of plate tectonics. According to this theory, earth's outer shell consists of about 20 plates that are essentially rigid, and **these plates move mostly laterally relative to one another like several large sheets of ice on a body of water**. The relative motion between points on different plates are, in some cases, as large as 150 mm/year, which is easily detectable using GPS and other modern day positioning techniques.

Given the fact that each tectonic plate is moving relative to the others, one may ask how crustal velocities may be expressed in "absolute" terms. The people responsible for ITRS currently address this dilemma by assuming that **the earth's surface, as a whole, does not move "on average" relative to Earth's interior**. Said differently, the ITRS developers assume that **the total angular momentum of earth's outer shell is zero**. Hence, the angular momentum associated with the motion of any one plate is compensated by the combined angular momentum associated with the motions of the remaining plates. Consequently, points on the North American plate generally move horizontally at measurable rates according to the ITRS definition of absolute motion. In particular, horizontal ITRF 96 velocities have magnitudes between 10 and 20 mm/year in the coterminous 48 states. Moreover, horizontal ITRF 96 velocities have even greater magnitudes in Alaska and Hawaii.

In contrast, the NAD 83 reference system addresses plate motion under the assumption that the North American plate, as a whole, does not move "on average" relative to earth's interior. Hence, points on the North American plate generally have no horizontal velocity relative to NAD 83 unless they are located near the plate's margin (California, Oregon, Washington, and Alaska) and/or they are affected by some other deformational process (volcanic/magmatic activity, postglacial rebound, etc.). The NAD 83 reference system, however, does make special accommodations for certain U.S. regions that are located completely on another plate. In Hawaii, for example, NAD 83 positional coordinates are defined as if the Pacific plate is not moving. This approach is convenient for people who are involved with positioning activities solely in Hawaii. This approach, however, introduces a layer of complexity for people who are involved in positioning points in Hawaii relative to points in

North America.

In the realm of crustal movement, it is inappropriate to specify positional coordinates without specifying the "epoch date" for these coordinates; that is, the date to which these coordinates correspond. Accordingly, ITRF96 positions are usually specified for the epoch date of 1 January 1997 (often denoted in units of years as 1997.0). To obtain positions for another time, t, people need to apply the formula

$$x(t) = x(1997.0) + v_x \cdot (t - 1997.0) \tag{8.30}$$

and similar formulas for $y(t)$ and $z(t)$. Here, $x(t)$ denotes the point's x-coordinate at time t, $x(1997.0)$ denotes the point's x-coordinate on 1 January 1997, and v_x denotes the x-component of the point's velocity.

8.7.2 Background of ITRF

One of the ultimate goals of space geodesy is to estimate point positions on the earth surface as accurately as possible. Meanwhile, **point positions are neither observable nor absolute quantities** and then have to be determined with respect to some reference. We refer to **"terrestrial reference system"** (TRS) as the mathematical object satisfying an ideal definition and in which point positions will be expressed. Nevertheless, the access to point positions requires some observational means allowing their link to the mathematical object. We therefore call a **"terrestrial reference frame"** (TRF), a physical materialization of the TRS, making use of observations derived from space geodesy techniques.

The distinction between "system" and "frame" is then subtle since the **former is rather invariable and inaccessible while the latter is accessible and perfectible**. The general concepts of reference systems and frames have been extensively discussed in the 1980s within the astronomical and geodetic communities. The use of space geodesy techniques since the 1980s has deeply improved positioning over the earth's surface: The uncertainties initially of decimetric level are now of centimetric, even millimetric level.

Nevertheless, each technique and each data analysis define and realize its own TRS. Therefore a multitude of TRF could exist, having systematic differences and bias when one is compared to another. This fact led the International Union of Geodesy and Geophysics (IUGG) and the International Association of Geodesy (IAG) to adopt a unique TRS, called the International Terrestrial Reference System (ITRS) for all earth science applications (Geodesist's Handbook, 1992). For the general description of the ITRS, see McCarthy (1996). The ITRS origin is defined by the center of mass of the whole earth, including oceans and the atmosphere. Its unit of length is the meter (SI) so that this scale is consistent with the geocentric coordinate time (TCG), in agreement with IAU/IUGG resolutions (1991), see the appendix of McCarthy (1992). The use of TCG ensures that all physical quantities take the same value in the terrestrial coordinate system as in the barycentric system (or in any other planetocentric system). The mean rate of the coordinate time TCG coincides with the mean rate of the proper time of an observer situated at the geocenter (with the earth removed), whereas the mean rate of the terrestrial time (TT) coincides with the mean rate of the proper time of an observer situated on the geoid. The two timescales differ in rate by (TCG - TT $\approx 0.7 \times 10^{-9}$). Its orientation is consistent with that of the Bureau International de l'Heure (BIH) at 1984.0. Its orientation time evolution is ensured by a **no-net-rotation condition** with regards to horizontal tectonic motions over the whole earth.

The basic idea of ITRF is to combine station positions (and velocities) computed by various analysis centers, using observations of space geodesy techniques, such as very long baseline VLBI, LLR and SLR, GPS, and DORIS. The combination method has proved its efficiency to produce a global reference frame which benefits from the strengths of all these different techniques.

The history of ITRF goes back to 1984, when for the first time a combined TRF (called BTS84), was established using station coordinates derived from VLBI, LLR, SLR, and Doppler/TRANSIT (the predecessor of GPS) observations. BTS84 was realized in the framework of the activities of BIH, being a coordinating center for the international project called Monitoring of Earth Rotation and Intercomparison of Techniques (MERIT). Three other successive BTS realizations were then achieved, ending with BTS87, when in 1988, the IERS was created by the IUGG and the International Astronomical Union (IAU).

Since the first ITRS realization, namely, ITRF88, nine other ITRF versions were established and published, each of which superseded and replaced its predecessor. Substantial improvements have since been constantly made in the data analysis strategy to achieve optimal combination for ITRF generation.

As the ITRF solutions are widely used in geodetic and geophysical applications, the ITRF2000 is intended to be an improved frame in terms of quality, network and datum definition. The ITRF2000 solution reflects the actual quality of space geodesy solutions, being free from any external constraints. It includes primary core stations observed by VLBI, LLR, SLR, GPS, and DORIS (usually used in previous ITRF versions) as well as regional GPS networks for its densification. To ensure its time evolution stability, the ITRF2000 orientation rate has been implemented on a selection of high quality geodetic sites.

Unlike the past ITRF versions where global long-term solutions were combined, the ITRF2005 uses as input data time series (weekly from satellite techniques and 24-h session-wise from Very Long Baseline Interferometry) of station positions and daily earth Orientation Parameters (EOPs). The advantage of using time series of station positions is that it allows to monitor station non-linear motion and discontinuities and to examine the temporal behavior of the frame physical parameters, namely the origin and the scale.

8.7.3 Background of ITRF Datum Definition

From a geodetic point of view, a TRS realization through a TRF, requires 14 parameters: three translations (origin), one scale factor, three rotations (orientation) and the seven corresponding rates. Given the fact that the 14 parameters should be viewed as relative values between two TRFs, their selection should correspond to the adopted TRS definition that one wants to satisfy. However, space geodesy observations do not carry all the necessary information to completely realize a TRS. While satellite techniques are sensitive to the earth center of mass (a natural TRF origin), VLBI is not (whose TRF origin is arbitrary defined through some constraints); the scale is dependent on the modeling of some physical parameters; and the TRF orientation (unobservable by any technique) is arbitrary or conventionally defined, through specific constraints.

The earth center of mass is the point with respect to which the orbit of dynamical techniques (LLR, SLR, GPS, DORIS) is referred. However, the center of mass is affected by various geodynamic processes of mass redistribution in the earth interior and surface envelopes, causing time variations of its position with respect to the earth surface. In other words, the geocentric motion of the tracking stations due to this effect, is known as "geocenter motion" and likely involves periodic and secular components. Presently, most of the analysis centers do not explicitly include this effect in their models. Therefore the origin of the resulting TRF coincides in practice with the position of the center of mass averaged over the period of the used observations. In practice, the dynamical techniques have currently limited abilities to accurately measure the geocenter motion.

When generating a combined TRF (such as the ITRF), different options could be adopted to specify explicitly the 14 datum parameters. While the seven parameters should be selected at a given epoch, the selection of scale and translation rates depends on the significance of the relative rates of the individual TRFs included in the combination.

The translation rates between satellite TRFs are heavily dependent on the network configuration, the orbit and the used observations. They may also be affected by the geocenter secular motion. The scale rate is influenced by station vertical motions and other modelling such as the troposphere, as well as technique-specific effects, such as VLBI, GPS, and DORIS antenna-related effects, and SLR station-dependent ranging biases. The orientation rate, although it could be arbitrarily chosen, should have a geophysical meaning due to tectonic plate motions.

Therefore a TRS realization should take into account motions and deformations of the earth's crust, simply because space geodesy observing sites are located on that deforming crust. Meanwhile, since space geodesy techniques make use of positions and/or motions of space objects, such as satellites and extra-galactic radio-sources in case of VLBI, consistency between Celestial Reference Frame (CRF), TRF, and EOP (connecting the two frames) should naturally be ensured for each individual (technique) TRS realization. Therefore the ITRF, being one of the three IERS global references, should be consistent with the ICRF and EOP series.

Moreover, fundamental equations of earth rotation theory are all related to a coordinate system in which any motion or deformation of the earth does not contribute to its net angular momentum. To achieve this goal, a generally accepted approach is the use of the Tisserand system of mean axes defined by minimizing the kinetic energy given by

$$T = \frac{1}{2} \int_C V^2 \mathrm{d}m \tag{8.31}$$

where $\mathrm{d}m$ is a mass element and V is its velocity. The most intuitive choice of the integration domain C is the earth's crust since, as it has already been mentioned, it "carries" our geodetic observing sites and the signals of its motion and deformation are contained in the observations.

One of the goals of defining a minimal kinetic energy T on the crust is to minimize its global motion that would affect the EOP. Meanwhile, deeper geodynamic processes such as mantle convection and the core-mantle boundary will still influence the earth rotation. Minimizing T leads to null linear p and angular h momentums:

$$p = \int_C V \mathrm{d}m = 0 \qquad (8.32)$$

$$h = \int_C X \times V \mathrm{d}m = 0 \qquad (8.33)$$

where X is the $\mathrm{d}m$ position vector.

Equations (8.32) and (8.33) theoretically define the TRF translation and rotation rates, respectively. The scale rate, although it is not foreseen in this approach, corresponds to a physical quantity estimated by space geodesy techniques.

Given the fact that satellite (dynamical) techniques estimate the geocenter motion, and so station positions are referred to the center of mass as a natural TRF origin, equation (8.32) is no longer needed. Meanwhile, equation (8.33) represents the definition of the no-net-rotation (NNR) condition over the crust.

A rigorous application of equation (8.33) requires knowledge of the crust density as well as its thickness which may vary spatially, especially between oceans and continents. However, a difficulty occurs when one wants to evaluate the integral of equation (8.33). A summation is then used instead, over a certain number of observing stations or tectonic plates representing the lithosphere. Each one of these two possibilities (stations or plates) leads to a different implementation of the NNR condition.

The approach of applying the NNR condition over observing stations requires careful investigation. The implied stations should satisfy minimum criteria, such as having a long enough observing period (say at least three years) and being distributed optimally over the whole earth's surface. Moreover, the use of equation (8.33) in this approach needs objective selection (or delimitation) of each area element associated with each station which should take its geophysical environment into account.

Since the ITRF94, we have started to use full variance matrices of the individual solutions incorporated in the global combination. At that time, the ITRF94 datum was achieved as follows: the origin (the three translation components), by a weighted mean of some SLR and GPS solutions; the scale, by a weighted mean of VLBI, SLR and GPS solutions, corrected by 0.7 ppb to meet the IUGG and IAU requirement to be in TCG (geocentric coordinate time) time frame instead of TT (terrestrial time) used by the analysis centers; the orientation was aligned to the ITRF92; and the time evolution was ensured by aligning the ITRF94 velocity field to the model NNR-NUVEL-1A over the seven rates of the transformation parameters.

The ITRF96 was then aligned to the ITRF94, and the ITRF97 was aligned to the ITRF96 using the 14 transformation parameters. Analysis of the individual solutions submitted to the ITRF2000 reveals that the ITRF97 exhibits a Z translation (and rate) with respect to most of the SLR solutions, as well as a scale factor.

It then became necessary to define the ITRF2000 datum to agree with the updated space geodesy data, and it was thus decided to adopt the following datum definition:

(1) The scale and its rate are defined by a weighted average of VLBI and the most consistent SLR solutions. Unlike the ITRF97 scale expressed in TCG-frame, that of the ITRF2000 is expressed in TT-frame, following the recommendations of the ITRF2000 Workshop (November 2000). This decision was adopted in order to satisfy space geodetic analysis centers contributed to ITRF2000 who use a time scale consistent with TT-frame.

(2) The translations and their rates are defined by a weighted average of the most consistent SLR solutions.

(3) The orientation is aligned to that of ITRF97 at epoch 1997.0 and its rate to be such that there is no-net-rotation rate with respect to NNR-NUVEL-1A. Note that the orientation as well as its rate are defined upon a selection of ITRF sites with high geodetic quality, satisfying the following criteria: continuously observed during at least 3 years; located on rigid parts of tectonic plates and far away from deforming zones; and velocity formal error (as result of the ITRF2000 combination) less than 3 mm/year; and velocity residuals less than 3 mm/year for at least three different solutions.

* * * * * * * * * * *

The ITRF2005 origin, scale, orientation and its time evolutions are specified as follows:

Origin: The ITRF2005 origin is defined in such a way that there are null translation parameters at epoch 2000.0 and null translation rates between the ITRF2005 and the ILRS SLR time series.

Scale: The ITRF2005 scale is defined in such a way that there are null scale factor at epoch 2000.0 and null scale rate between the ITRF2005 and the IVS VLBI time series. VLBI scale selection to define that of ITRF2005 is justified by the availability of the full VLBI history of observations (26 years versus 13 for SLR) embedded in the submitted time series and the non-linear behavior (discontinuities) observed in the ILRS scale.

Orientation: The ITRF2005 orientation is defined in such a way that there are null rotation parameters at epoch 2000.0 and null rotation rates between the ITRF2005 and ITRF2000. These two conditions are applied over a core set of 70 stations.

The ITRF2005 origin is defined in such a way that it has zero translations and translation rates with respect to the earth center of mass, averaged by the Satellite Laser Ranging (SLR) time series spanning 13 years of observations. Its scale is defined by nullifying the scale and its rate with respect to the Very Long Baseline Interferometry (VLBI) time series spanning 26 years of observations. The ITRF2005 orientation (at epoch 2000.0) and its rate are aligned to the ITRF2000 using 70 stations of high geodetic quality. The estimated level of consistency of the ITRF2005 origin (at epoch 2000.0) and its rate with respect to the ITRF2000 is respectively 0.1, 0.8, 5.8 mm and 0.2, 0.1, 1.8 mm/year along the X, Y and Z-axis.

* * * * * * * * * * *

Following the same strategy initiated with the ITRF2005 release, the ITRF2008 is a refined solution based on reprocessed solutions of four space geodesy techniques: VLBI, SLR, GPS and DORIS, spanning 29, 26, 12.5 and 16 years of observations, respectively.

The ITRF2008 is composed of 934 stations located at 580 sites, with an imbalanced distribution between the northern (463 sites) and the southern hemisphere (117 sites).

There are in total 105 co-location sites; 91 of these have local ties available for the ITRF2008 combination. Note that, unfortunately, not all these co-located instruments are currently operating. For instance, among the 6 sites having 4 techniques, only two are currently fully operational: Hartebeesthoek, South Africa and Greenbelt, MD, USA.

The ITRF2008 is specified by the following frame parameters:

(1) Origin: The ITRF2008 origin is defined in such a way that there are zero translation parameters at epoch 2005.0 and zero translation rates with respect to the ILRS SLR time series.

(2) Scale: The scale of the ITRF2008 is defined in such a way that there is a zero scale factor at epoch 2005.0 and a zero scale rate with respect to the mean scale and scale rate of VLBI and SLR time series.

(3) Orientation: The ITRF2008 orientation is defined in such a way that there are zero rotation parameters at epoch 2005.0 and zero rotation rates between ITRF2008 and ITRF2005. These two conditions are applied over a set of 179 reference stations located at 131 sites.

Vocabulary

least squares /liːst skwɛəs/ 最小二乘

no-net-rotation (NNR) condition /nəu net rəuˈteiʃən kənˈdiʃən/ 无绝对旋转条件

References

ALTAMIMI Z, SILLARD P, BOUCHER C. 2002. ITRF2000: A new release of the International Terrestrial Reference Frame for earth science applications[J]. J. Geophys. Res., 107(B10): 2214, doi: 10.1029/2001JB000561.

ALTAMIMI Z, COLLILIEUX X, LEGRAND J, et al. 2007. ITRF2005: A new release of the International Terrestrial Reference Frame based on time series of station positions and Earth Orientation Parameters[J]. J. Geophys. Res., 112: B09401, doi: 10.1029/2007JB004949.

Encyclopaedia Britannica. 2009[2009-10-01]. Encyclopaedia Britannica 2009 Student and Home Edition[CD]. Chicago: Encyclopaedia Britannica.

HOFMANN-WELLENHOF B, MORITZ H. 2005. Physical Geodesy [M]. Springer Wien NewYork.

IAG. 2008. Geode sist's Handbook[J]. J. Geod, 82: 662-846.

MCCARTHY, D D (ed.). 1992. IERS Standards (1992) [M]. IERS Technical Note 13, Observatoire de Paris, Paris.

MCCARTHY, D D (ed.). 1996. IERS Conventions (1996) [M]. ERS Technical Note 21, Observatoire de Paris, Paris.

MCCARTHY, D D, PETIT G (eds.). 2004. IERS Conventions (2003)[M]. IERS Technical Note 32, BKG, Frankfurt am Main.

MORITZ H. 2000. Geodetic Reference System 1980[J]. J, Geod, 74: 128-133.

PETIT G, LUZUM B (eds.). 2010. IERS Conventions (2010) [M]. IERS Technical Note 36. Frankfurt am Main.

RUMMEL R. 2010. The interdisciplinary role of space geodesy-Revisited [J]. Journal of Geodynamics, 49: 112-115.

TORGE W. 2001. Geodesy[M]. 3rd edition. Berlin, New York: de Gruyter.

U.S. Defense Mapping Agency. 1983. Geodesy for the layman[R]. TR 80-003.

U.S. National Geo-intelligence Agency. 2002. Addendum to NIMA TR 8350.2: Implementation of the World Geodetic System 1984 (WGS-84) Reference Frame G1150[R].

U.S. National Imagery and Mapping Agency. 1997. Department of Defense World Geodetic System 1984: Its Definition and Relationships with Local Geodetic Systems[R]. Third Edition. TR 8350.2.

Vocabulary

altimetry /æl'timitri/ *n.* 测高学, 高度测量法(以海平面为基准)
astronomical system /əs'trɔnəmik 'sistəm/ 天文系统
azimuth /'æziməθ/ *n.* 方位角
baseline /'beislain/ *n.* 基线
Cartesian coordinate system /kɑː'tiːzjən kəu'ɔːdineit 'sistəm/ 笛卡儿坐标系(直角坐标系)
celestial /sə'lestʃəl/ *adj.* 天的, 天空的
centrifugal /sen'trifjəgəl – 'trifə –/ *adj.* 离心的
conformal projection /kən'fɔːməl prə'dʒekʃən/ 正形投影
control network /kən'trəul 'netwəːk/ 控制网
Conventional International Origin /kən'venʃənl ˌintəˈnæʃənəl 'ɔridʒin/ 国际协议原点
convergence of meridian /kən'vəːdʒəns ɔv məˈridiːən/ 子午线收敛角
crustal movement /'krʌstəl 'muːvmənt/ 地壳运动
curvature /'kəːvəˌtʃuə, – tʃə/ *n.* 曲率, 弯曲
deflection of the vertical /di'flekʃən ɔv ðə 'vəːtikəl / 垂线偏差
development method /di'veləpmənt 'meθəd/ 展平法
direction measurement /di'rekʃən 'meʒəmənt/ 方向测量
distance measurement /'distəns 'meʒəmənt/ 距离测量
dynamic height / dai'næmik hait/ 力高
earth ellipsoid /əːθ i'lipsɔid/ 地球椭球
earth orientation parameters (EOP) /əːθ ˌɔːrien'teiʃən pə'ræmitəs/ 地球定向参数
earth rotation /əːθ rəu'teiʃən/ 地球自转
ellipsoid /i'lipsɔid/ *n.* 椭球
equatorial /ˌiːkwə'tɔːriːəl/ *adj.* 赤道的
equipotential /ˌiːkwipəu'tenʃəl/ *adj.* 等位的
figure of the earth /'figə ɔv ði əːθ/ 地球形状
flattening /'flætniŋ/ *n.* 扁率
Fundamental Catalogue /ˌfʌndə'mentəl 'kætəlɔg/ 基本星表
geocenter /ˌdʒiːəu'sentə/ *n.* 地球质量中心
geocentric system /ˌdʒiːəu'sentrik 'sistəm/ 地心体系
geodesic /ˌdʒiːəu'desik/ *adj.* 大地测量的, 测量的; *n.* 大地线
geodesy /dʒiː'ɔdisi/ *n.* 大地测量学
geodetic /ˌdʒiːə'detik/ *adj.* 大地测量的, 测量的
geodetic datum /ˌdʒiːə'detik 'deitəm/ 大地基准
geodetic origin /ˌdʒiːə'detik 'ɔridʒin/ 大地原点

geodetic reference system (GRS) /ˌdʒiːə'detik 'refrəns 'sistəm/ 大地测量参考系
geoid /'dʒiːɔid/ n. 大地水准面
geoid height /'dʒiːɔid hait/ 大地水准面高
geopotential number /ˌdʒiːəupə'tenʃəl 'nʌmbə/ 地球位数
gravimeter /grə'vimitə/ n. 重力仪
gravity field /'græviti fiːld/ 重力场
Greenwich meridian /'grinidʒ mə'ridiːən/ 格林尼治子午线
horizontal control network /ˌhɔri'zɔntəl kən'trəul 'netwəːk/ 平面控制网
horizontal direction /ˌhɔri'zɔntəl di'rekʃən/ 水平方向
interferometry /ˌintəfiə'rɔmitri/ n. 干涉测量术
isometric latitude /ˌaisəu'metrik 'lætitjuːd/ 等量纬度
Laplace azimuth /lɑː'plɑːs 'æziməθ/ 拉普拉斯方位角
latitude /'lætitjuːd/ n. 纬度
least squares /liːst skwɛəs/ 最小二乘
Legendre functions /lə'ʒɑːŋdə 'fʌŋkʃənz/ 勒让德函数
leveling /'levəliŋ/ n. 水准测量
leveling network /'levəliŋ 'netwəːk/ 水准网
level surface /'levl 'səːfis/ 水准面
longitude /'lɔndʒitjuːd/ n. 经度
mean sea level (MSL) /miːn siː 'levl/ 平均海(水)面
meridian /mə'ridiːən/ n. 子午圈, 子午线
network /'netwəːk/ n. (控制)网
no-net-rotation (NNR) condition /nəu net rəu'teiʃən kən'diʃən/ 无绝对旋转条件
normal gravity /'nɔːməl 'græviti/ 正常重力
normal height /'nɔːməl hait/ 正常高
normal section /'nɔːməl 'sekʃən/ 法截面
nutation /njuː'teiʃən/ n. 章动
orientation /ˌɔːrien'teiʃən/ n. 方向, 定向
orthometric height /ɔː'θɔmitrik hait/ 正高
orthomorphic /ˌɔːθə'mɔːfik/ adj. 正形的
plate tectonics /pleit tek'tɔniks/ 板块构造学
plumb line /plʌm lain/ 铅垂线
polar motion /'pəulə 'məuʃən/ 极移
polar radius of curvature /'pəulə 'reidjəs ɔv 'kəːvəˌtʃuə/ 极曲率半径
precession /pri'seʃən/ n. 岁差, 进动
prime vertical /praim 'vəːtikəl/ 卯酉圈
quasi-geoid /ˌkwəːzi(ː) 'dʒiːɔid/ n. 似大地水准面
reference ellipsoid /'refrəns i'lipsɔid/ 参考椭球
rotational ellipsoid /rəu'teiʃənəl i'lipsɔid/ 旋转椭球

scale factor /skeil ˈfæktə/ 长度比
sidereal time / saiˈdiəriːəl taim/ 恒星时
spherical excess /ˈsfiərikəl ˈekses/ 球面角盈
spherical harmonic expansion /ˈsfiərikəl hɑːˈmɔnik iksˈpænʃən/ 球谐展开
spheroid /ˈsfiərɔid/ n. 球状体, 旋转椭球
terrestrial reference frame / təˈrestriːəl ˈrefrəns freim/ 地球参考框架
theodolite /θiˈɔdəlait/ n. 经纬仪
total station /ˈtəutəl ˈsteiʃən/ 全站仪
traverse /ˈtrævəs/ n. 导线
triangulation /traiˌæŋgjuˈleiʃən/ n. 三角测量
trigonometric leveling /ˌtrigənəˈmetrik ˈlevəliŋ/ 三角高程测量
trilateration /traiˌlætəˈreiʃən/ n. 三边测量
universal time /ˌjuːniˈvəːsəl taim/ 世界时
vernal equinox /ˈvəːnəl ˈiːkwinɔks/ 春分点
vertical datum /ˈvəːtikəl ˈdeitəm/ 高程基准
zenith /ˈziːniθ/ n. 天顶
zenith distance /ˈziːniθ ˈdistəns/ 天顶距

参 考 译 文

1 概 论

1.1 大地测量学的定义

大地测量学是研究地球和其他天体的测量与表示(几何、物理、时变)的学科。

****

按照 F. R. Helmert(1880)的经典定义,大地测量学是"地球表面测量与制图的科学"。直到今天,这个定义仍然有效,包括地球外部重力场的确定,以及海底地形的确定。按照这个定义,大地测量学既可包含于地球科学,亦可归为工程科学。

****

Webster 词典将大地测量学定义为"应用数学的分支,通过观测和测量,确定点位的精确位置、地球表面大片区域的形状和范围、地球的形状和大小,以及地球重力的变化",是基础数学和物理概念中几个常见方面的特殊应用。实际上,大地测量学用到了数学、天文学和物理学中的原理,并在现代工程和技术水平下应用这些原理。

****

关于地球的精确形状,及其确定与意义的科学学科。人造卫星出现之前,所有的大地测量工作都是基于地面测量的,采用大地坐标系(用来研究弯曲表面的几何属性),利用三角测量方法实现。现在,可以利用卫星结合地面系统,精化对地球形状和大小的知识,有时将这项工作称为卫星大地测量学。

1.2 大地测量学的分类

大地测量学可分为全球大地测量学、控制测量学和平面测量学。全球大地测量学负责确定包含整个外部重力场的地球形状。控制测量学通过一批足够数量控制点的坐标来定义一个国家的表面。在这一基础工作中,必须考虑到地球的总体弯曲。在平面测量学(地形测量、工程测量、地籍测量)中,将获取陆地表面的细部,一般把水平面作为参考面。

全球大地测量学、控制测量学和平面测量学之间存在着密切的联系。控制测量需要全球大地测量所确定的参数,控制测量的结果又为全球大地测量提供信息。然后,平面测量通常与控制测量中的控制点相连接,以便满足国家系列地图测绘的需要,以及地产地籍造册的需要。

1.3 大地测量学的任务

工程任务:
应用大地测量学的科学结果,开展地球表面精确、可靠制图所需要的测量和计算。
科学任务:
为了确定地球的形状和大小,联合其他科学,研究地球重力场,以及在一定程度上研究地球内部结构。

1.4 历史简介

约 150 年前,才将大地测量学作为一门独立学科,Baeyer 的关于地球形状和大小的备忘录 "Über die Größe und Figur der Erde"(1861),可以看作是起点。在这之前,尽管著名的科学家如牛顿、拉普拉斯、高斯和贝塞尔,已经完成了重要的大地测量工作,但并没有把他们的工作作为大地测量,他们也没有把自己当作大地测量学家。Baeyer 首倡将当时覆盖中部欧洲的三角网和水准网扩展并统一,后来这项工作扩展至整个欧洲。再后来,这项工作变成一项国际性的工作,目标是确定地球的整体形状。这是科学上首次国际性工程之一,是现今国际大地测量联合会(IAG)的根源。

2007 年,人类庆祝迈入太空时代 50 周年。1957 年 10 月 4 日,随着 Sputnik1 号的发射(之后不久发射了 Sputnik2 号),开启了现代空间时代。这两颗卫星对大地测量学产生了重要的影响。瞬间,致力于确定地球形状一百多年的大地测量学的勤奋工作,很大一部分变得过时。通过测量卫星轨道的进动,比用经典天文大地法,确定的地球扁率要准确得多。卫星为大地测量开启了崭新的视野,据我所知,没有其他学科比大地测量学受空间技术的影响更为深刻。通过空间技术,定位、重力场确定、地球自转监测和测绘遥感等工作可以完成的更加准确、完整和有效。大地测量学变成了真正的全球三维测量。海洋,在经典时代的"未知领域",随着卫星的出现,成为可以从事大量大地测量活动的区域;由于大气折射,经典大地测量技术不能够精确测量天顶距,而从太空中,地球表面的垂直角可以像水平分量测得一样准确。空间大地测量学发展迅速,影响深远。随着大地测量空间技术的迅速发展,地球科学越来越关注大地测量的工作。

1.5 空间大地测量学的近来发展

近年来,地球科学的重点转移到气候变化和地球系统科学。人类对以下问题的认识不断加深,人类需要更好地理解地球是一个系统,更好地理解太阳辐射是地球的动力,更好地理解热反向辐射及其如何受大气化学成分的极微小变化的影响,最后但并非次要的一点,更好地理解人类的影响。在起草政府间气候变化专门委员会的最近一份报告时(2007),一个基本的缺陷变得特别明显,这个问题在《科学》和《自然》杂志上的几篇文章中都有论述:明显缺乏观测支持。空间大地测量通过测量地球系统中的质量和能量迁移过程,能够为全球变化研究提供新的、重要的、独一无二的数据。Ben Chao(2003)写道:"经过三十年的发展和三个数量级的提高,空间大地测量已经准备好大显身手,观测地球系统中发生的总质量迁移,包括从高层大气到地核内部的质量迁移。这样,空间大地测量就变成一种新的遥感工具,以越来越高的敏感度和分辨率用于监测气候和地球物理变化"。可以断言,融合了几何定位、地球自转、重力场与大地水准面的大地测量学,能够为地球系统研究提供"度量体系"。在这一背景之下,全球大地测量观测系统(GGOS)的建立是在恰当的时间做出的正确举措。GGOS 的基本概念很简单,Rothacher 给出的图示很好地描述了这一概念(见图 1.1)。

GGOS 联合了大地测量学的三个基本任务:测量地球的形状,地球自转,地球重力场和大地水准面。目标是,在一个统一的地固参考系中以 10^{-9} 的精度水平实现这三项任务,并保持系统数十年的稳定。这一高难的需求来自哪里呢?在地球科学中,人们通常仅以百分之几的精度处理待估量,全球变化参数是很小的,其时变量变化缓慢,数值更小。一般而言,它们不能直接通过观测得到,而必须从几种观测系统和模型的联合处理中导出。为了能够把它们作为全球过程分析,必须相对于地球的尺寸来衡量。举例说明,任意一个验潮站的海平面,由于潮汐和风暴,可能会有几米的变化,因此,以几个 mm 的精度测量海平面变化需要在这个特定测站上的相对精度达到 10^{-3}。通过卫星测高,GPS 等卫星系统,区域海平面监测可以转换到全球监测系统,只有在这时,才能够从当地验潮记录中发现全球过程变化。为了在全球范围内实现卫星系统达到厘米或毫米级精度,定轨和测高必须要以 1 ppb 的相对精度测定。

为了满足 GGOS 设定的目标,一系列相当基础的大地测量问题必须得到解决。大地测量学的三大任务,几何定位、地球自转和重力场,必须以毫米精度在统一的地固参考系中表示,还要保持参考框架数十年的稳定。这要求空间部分和地面部分像一个同质的实体一样运行,所有的观测如同是在围绕地球的同一个观测台站上完成。每个观测量包含了多种效应的叠加,包括与电离层、大气、海洋、冰盖和固体地球等有关的效应。为了将这些观测量用于地球系统研究,必须开发各种策略,通过分析这些效应的空间、时间和谱特征,来分离和量化它们。对卫星的观测量代表沿卫星轨道的时间序列,通过地球自转和卫星轨道根数的选择,这些时间序列与地球的空间和时变采样建立联系,重构地球物理时间和空间现象提出了一个复杂的叠加和反演问题。当前,根据 GRACE 重力观测资料,对全球水循环的研究,或者是对格陵兰岛和南极洲冰质量平衡的研究,都属于这类问题。将地面和空载数据包括进来,如表面负载、海底压力、验潮、重力测量或测高等,可能会很有帮助,但是,这一步并不容易,因为地面测量受到当地影响,所表现出的谱敏感性与卫星观测所表现出来的完全不同。也许分离地球物理现象最有效的手段是将先验信息包括进来,例如固体潮和海潮模型,大气、海洋、冰、水文、冰均衡调整等,当然,这些先验信息对观测系统的所有技术都要一致地引入。当前,面向这些目标的重要工作正在开展,可以看到大地测量技术在各种地球学科中都得到广泛得多的应用。

魏格纳(Wegener)经过多年努力,将一个几乎单学科的工程发展成一个区域性的多学科活动,联合了大地测量和非大地测量中的许多测量技术,涉及跟工程目标有关的所有地学学科。因此,这是 GGOS 可以在全球尺度上运行的一个极好示例。

2 参考系统

2.1 时间系统

在这本书中,把时间当作绝对的、与空间独立的。现实中,现在通常的时间测量和卫星观测,精确到需要应用狭义和广义相对论来建立精确模型,以满足实用。

时间测量需要一个周期过程,一个计数器(对周期计数)和一个开始计数的起点。此外,为了能够在不同的地点保持同样的时间,必须要有时间传递手段。有大量的自然"时钟",能产生很稳定的周期振荡:地球绕太阳的轨道,月亮绕地球的轨道,以及地球自转。它们的基本周期即年、月、日与自然过程紧密相连,例如影响人类生活条件的季节,并且这些周期定义了人类生命的基本结构。从这些基本周期,演绎出基本的长期计数结构,即历法(现在用的是自1582年采用的格里高利历)。在科学工作中,相比带有长度变化的月或年的这种复杂计数结构,连续计数更可取。为了这个目的,发明了儒略日(JD),每世纪有36 525日。

采用的参考儒略日为

$$J2000.0 = 2000 \text{ Jan } 1.5 = \text{January } 1, 2000 \text{ at } 12h$$

它是

$$JD\ 2451545.0.$$

很长一段时间,自然周期天,甚至月亮的旋转,都比任何人造时钟稳定性好,只有当石英和原子振荡器出现后,人造时钟才在精度和稳定性上超过自然时钟。当前,单位秒长的定义是基于铯钟振荡周期的。1984年,引入原子时(TAI)作为正式的国际时间。TAI相对于地球动力学时(TDT)有一个恒定偏差32.184 s。后者是根据行星运动模型导出的,以相对论为基础。TAI相对于GPS时(GPST)有19 s的偏差。

民用时间与日夜的更替,即与太阳的升起和下落有关。由于太阳视运动的不均匀,建立了某种模型,也就是平太阳运动,它参考于格林尼治子午圈,称为世界时(UT),由世界时演绎出标准区时。地球自转——从而是世界时——相对于TAI表现出漂移和小的不规则波动(日长LOD变化),为了回避这一点,建立了协调世界时(UTC);一方面保持与TAI的同步,另一方面,通过有规律的调整(闰秒),保持与地球实际角速度在很小的范围内同步。实际的、未经改正的世界时表示为UT1,代表了旋转地球的实际相位角,UT1-UTC之差以月表形式提供,在广播时号中以编码信息形式提供。当UT1-UTC之差超过0.9 s时,引入跳秒。UT0完善了世界时系统,它包含了极移引起的所有自转变化。

最后,恒星时是地球子午圈(随地球自转)相对于春分点的角度。最重要的恒星时类型为格林尼治平恒星时GMST和格林尼治真恒星时GAST。

GMST加入章动引起的波动改正,即

$$\Theta = \bar{\Theta} + \Delta n = \bar{\Theta} + \Delta\psi \cdot \cos\varepsilon + 0.002\ 64'' \cdot \sin\Omega + 0.000\ 063'' \cdot \sin2\Omega \quad (2.1)$$

式中,Ω为月亮的平升交点。在将地固系转换至空固系时需要GAST。

由于恒星时是相对于春分点起算的,而世界时是一种太阳时,是从视太阳通过格林尼治子午圈量算的,因而年的长度相差一天:回归年,其长度用太阳日计算为365.242 20天,用恒星日计算为366.242 20天。

当把UT1转换为GMST时必须考虑这一差别

$$GMST = UT1 + \alpha(\Theta) - 12^h \quad (2.2)$$

式中,太阳的升交点为

$$\alpha(\Theta) = 12^h + (24\ 110.548\ 40^s + 8\ 640.812\ 866^s \cdot t + 0.093\ 104^s \cdot t^2 - 6.2^s \cdot 10^{-6} \cdot t^3) \quad (2.3)$$

且

$$t = (T - J2000.0)/36\ 525.0 \quad (2.4)$$

式中,T为历元的儒略日,J2000.0 = JD2 451 545.0。

2.2 国际地球自转服务(IERS)

IERS 负责建立和维持协议天球和地球参考框架,这些框架是 IAU 和 IUGG 建议的参考系的实现。IERS 还负责确定地球随时间变化的定向参数,这些参数可以用来建立这两种框架的联系。

IERS 是由 IAU 和 IUGG 设立的,自 1988 年 1 月 1 日运行。负责全球天文和大地台站网络(1996 年约 300 个台站)观测资料的收集、分析、建模,这些台站有的是连续运行的,有的是按某一时间间隔运行的。观测技术包括 VLBI、LLR、SLR、GPS、DORIS。

不同类型的观测资料在相应的 IERS 协调中心进行处理,然后由 IERS 中央局联合平差,结果包括河外射电源和地面站的位置(坐标),地球定向参数(EOP),以及其他信息。至于 EOP,VLBI 提供关于岁差、章动、极移和 UT1 的信息;卫星技术提供对 UT 的每日内插值,以及确定极移。结果通过公报、年报和技术备忘录公布。观测资料的计算是基于《IERS 规范》的,该规范与 IAU 和 IUGG/IAG 关于参考系的建议一致。

2.3 天球参考系

当描述地球和其他天体在空间的运动时,包括人造卫星天体,需要用到惯性系。惯性系可以用牛顿运动定律来描述,惯性系是指处于静止或无旋转地匀速直线运动状态。空固系(天球参考系)代表了对惯性系的一种近似,可以通过合理的协议来定义,即协议惯性系(CIS)。这种系统的坐标框架由球面天文学来建立,这一框架的空间定向随时间而变化,因此,需要建立这种变化的模型。当前 ICRF 是天球参考系的实现。

2.3.1 球面天文学中的赤道坐标系

天球参考系的坐标由球面天文学中的赤道坐标系来定义。引入原点在地球质心(地心)的三维直角坐标系,Z 轴与地球自转轴重合,X 和 Y 轴构成赤道平面,X 轴指向春分点,Y 轴构成右手系(见图 2.1)。

绕地球画出单位球(天球),自转轴与球分别交于北天极和南天极 P_N 和 P_S。包含天极且与天球赤道垂直的大圆,称为时圈,与赤道平行的小圆称为天球纬圈。

赤经 α 是在赤道面量算的过春分点的时圈和过天体 S 的时圈之间的夹角,从春分点开始逆时针方向量算。赤纬 δ 是在时圈内量算的赤道面和 OS 线之间的夹角(从赤道向 P_N 为正,向 P_S 为负)。

天体 S 的位置既可以用直角坐标 XYZ 来表示,也可以用球坐标 α, δ, r(r = 从原点量起的距离)来表示。其相互转换关系为

$$r = \begin{pmatrix} X \\ Y \\ Z \end{pmatrix} = r \begin{pmatrix} \cos\alpha\cos\delta \\ \sin\alpha\cos\delta \\ \sin\delta \end{pmatrix} \tag{2.5}$$

在大地测量学中,对于恒星和河外射电源,只有方向是重要的,故取 $r = 1$,α 和 δ 描述的是单位球上的天体 S 的位置,也可以用赤道和时圈上的相应弧长来表示。

引入观测者所在的当地子午面,即由当地垂线(铅垂线方向)和自转轴所张成的面,自转轴需从地心平移至站心。天顶 Z 是垂线与单位球的交点,天球子午面是过 Z 和南北极的大圆(见图2.2)。时角 h 是过 Z 的天球子午面和过 S 的时圈之间的夹角,在赤道面内从上子午圈向西量取。由于地球自转,时角系统(h,δ)与时间有关。相对于 α,δ 系统,h,δ 系统是绕极轴旋转的,旋转角为恒星时角 LAST。存在以下关系

$$\text{LAST} = h + \alpha \tag{2.6}$$

该关系用于定时。

2.3.2 岁差和章动

前面作为坐标系 Z 轴引入的地球自转轴,相对于空间的定向是随着时间变化的。因此,天体位置(h,δ)是变化的,叠加了一系列长周期和短周期的效应。

日月岁差是由太阳和月亮对赤道隆起部分的吸引而引起的一种长周期效应,所产生的力偶(矩)要把赤道平面拉回到黄道平面(见图2.3),再加上地球的自转力矩,地球自转轴的运动构成了一个顶角为 23.5°(即黄赤交角)、绕北黄极的圆锥,春分点沿黄极按顺时针方向以 50.3″/a 的速度移动,周期约为 25 800 a。行星的引力造成地球轨道的缓慢移位,从而造成春分点沿赤道的附加位移和 ε 的变化,这称为行星岁差。日月岁差和行星岁差的和称为总岁差。

岁差还叠加了一系列短周期效应,称为章动,周期从 5 天到 18.6 年不等。这些周期主要是由于月球轨道相对于黄道轨道的倾角(约为 5°)变化引起的;其他还包括以半年和半月为周期的分量,是由太阳和月球在地球南北半球之间的振荡引起的。

岁差和章动可以用月球、太阳和行星的历书表示成时间函数的模型。IAU(1976)章动理论提供了三个与时间有关的欧拉旋转角,用来把天体位置归算为一般参考位置。在参考历元 J2000.0,有基本常数"黄经总岁差"(5 029.096 5″/世纪)和"黄赤交角"(23°26′21.412″)。

IAU(1980)章动理论把章动效应描述为绕岁差圆锥的旋转,真极与平极的偏差表示为两个与时间有关参数的模型。需指出,地球指的是椭球形的、自转的、弹性的、无海洋的形体,且包含固体内核和液态外核。在历元 J2000.0,章动常数为 9.202 5″。

IAU 的岁差和章动模型定义了国际天球参考框架的参考极(天球历书极 CEP)。相对于空固或地固坐标系,CEP 不包含周日或近周日章动项(幅度 < 0.001″),也作为瞬时赤道系的极。

IAU 的岁差和章动模型在 5 到 7 天时间分辨率上给出了 0.001″ 的精度。基于近来的 VLBI 和 LLR 数据,IERS 已发展了新的改进模型。已经发现了天极与 CEP 的更大的偏差(<0.02″);由 IERS 定期公布。

天体的瞬时位置称为在历元 t 的真位置。顾及章动,可以得到历元 t 的平位置,平位置是相对于平天赤道和平春分点的。如果还顾及到岁差,可以得到在参考历元 J2000.0 的平位置。

2.3.3 国际天球参考框架

IAU 推荐的国际天球参考系统(ICRS),是基于广义相对论的,其时间坐标是由国际原子时定义的。ICRS 接近于空固协议惯性系(CIS),原点在太阳系质心。并假定该系统不存在整体旋转,这意味着定义源或者是没有自行(与天球相切的空间运动分量),或者是运动可以建模。坐标轴由 IAU 岁差和章动模型给出的天球参考极和春分点定义。它们是通过到地球外

基准天体的方向实现的恒星或射电源 CIS。

恒星系统是基于 FK5 基本星表恒星的。给出了 1535 颗基本星在历元 J2000.0 的平坐标 (α,δ) 和自行（一般 $<1''$/年），精度分别是 $\pm 0.01 \sim 0.03''$/世纪和 $\pm 0.05''$/世纪。FK5 的补充星表包含了视星等直到 9.5 的附加恒星。历元 J2000.0 的平赤道和平春分点由 FK5 星表实现，精度为 $\pm 0.05''$。由于大气折射的影响，地基天体测量很难再提高这一精度。

天文空间观测计划显著改善了恒星 CIS 的实现。依巴谷（HIPPARCOS）天体测量卫星（ESA,1989—1993）通过测量覆盖全天区的约 100 000 颗（直到视星等 9 等）恒星间的大角度，用来建立一个网络。这样建立的参考框架给出的自行精度为 $\pm 0.001''$/a 和 $\pm 0.000\,5''$/a。根据改进的 FK5 数据和依巴谷结果，建立了含较少恒星（340 颗"天文测量上极好"的恒星）的 FK6 星表，与依巴谷星表相比，对自行有了很好地完善。未来的天体测量空间使命将采用光学干涉技术，因而可将坐标精度提高到 $\pm 0.000\,01''$。

射电源系统基于河外射电源（类星体和其他致密源）。1997 年，IAU 将射电源系统引入 ICRS，并且从 1998 年起取代了以前的恒星系统（FK5）。由于距离遥远（>15 亿光年），这些射电源没有表现出可观测得到的自行。该系统由 ICRF 实现，该框架由 IERS 建立和维持。ICRF 包括 600 多个源的坐标（赤道系统，J2000.0 历元），其中约 200 个为观测质量良好的"定义源"，100 多个用于加密以及与恒星星表系的联结。由于望远镜都集中在北半球，南半球没有覆盖。射电源坐标由射电天文学测定，平均精度优于 $\pm 0.001''$，最精密的观测源精度可达 $\pm 0.000\,3''$。

2.4 地球参考系

引入地固参考系，用于地球表面或接近地面的定位和导航，用来描述地球重力场以及其他物理参数。它被定义为三维地心坐标系。这一系统的定向，相对于固体地球体以及天球参考系随着时间而变化。该系统由 IERS 提供的 ITRF 实现。

2.4.1 全球地固地心系

空间直角坐标 X,Y,Z 的地固（即，随地球自转）系，是基本的地球坐标系，其原点在地球质心（地心），定义为包含水圈和大气在内的整个地球的质心；Z 轴指向协议"平"地（北）极，"平"赤道面与 Z 轴垂直且包含 X 轴和 Y 轴。地球自转相对于地球本体是随时间而变化的，因此必须要引入"平"自转轴和"平"赤道面。XZ 平面为格林尼治协议"平均"子午面，由平均自转轴和格林尼治零子午圈张成，世界时 UT 就是参照该面的。Z 轴和 X 轴通过地面"基准"站的坐标间接实现，Y 轴的定向与 XZ 轴构成右手系。

2.4.2 极移、日长、地心变化

地球自转可以用指向瞬时自转轴北极方向的矢量和角速度 ω 来描述。

旋转矢量的大小和方向随时间而变化，是由天文和地球物理过程造成的。这些过程包括太阳及月球引力的变化，大气层、水圈、固体地球和液内核的质量重新分配。从性质上讲，这些变化有长期的、周期或近似周期的以及不规则的等种类。

从地固参考系来看，极移（抖动）是自转轴相对于地壳的运动，对地球表面的测站坐标和

重力矢量产生直接影响。极移包含以下分量：

(1) 周期约为 435 天(钱德勒周期)的自由振荡,幅度从 0.1″至 0.2″,从北极看为逆时针方向。钱德勒抖动是由于地球自转轴与主惯性轴不完全重合引起的。

(2) 钱德勒抖动叠加有年度振荡,是由大气和水质量的季节性位移造成的,它和钱德勒抖动沿相同方向运动,幅度从 0.05″到 0.1″。

(3) 通过 100 多年的观测资料发现了地极的长期运动,这一运动是一种向 80°W 子午圈方向、速率为 0.003″/a 的不规则漂移。长期运动主要是由于极地冰川融化和大尺度构造运动引起的,经过数个地质时期可以产生很大的量:即极游动。

(4) 在从几天到数年不等的时间尺度上,还有更多不规则变化,幅度最大可达 0.02″。主要是由大气质量重新分配引起的,海洋体积变化、地下水变化和地震也会引起地极变化。

这些分量的叠加产生了一条轻微扰动的瞬时极螺旋线,伴随着缓慢移动的平极。相对于平极每年产生 < 0.3″的偏差,在地球表面相当于 9 m。

描述地极相对于固体地球实际位置的参考是 IERS 参考极,它与国际协议原点 CIO 在 ±0.03″以内是一致的,CIO 是由 1900.0 和 1906.0 期间确定的北极平均位置定义的。瞬时极(天球历书极)相对于参考极的位置由直角坐标 x_P, y_P 表示,该坐标定义在与北极相切的平面内,x_P 轴指向格林尼治平子午圈(与以前的 BIH 零子午圈一致)方向,y_P 轴指向 90°W 子午圈方向。这些平面坐标通常表示成单位球上的球面边长形式(以弧秒为单位)。

从地球上监测的地球自转角速度 ω 随时间的变化,相对变化会超过 10^{-8},相当于一天几个毫秒。这种变化一般用相对于 86 400 秒超出的时间来表示,称为日长(LOD),通过比较天文测定的时间,即世界时 UT1,和均匀时间尺度 TAI 或 UTC 得到日长。

已经观测到的日长变化包括如下分量：

(1) 地球自转角速度的长期减慢,主要由潮汐摩擦所致,使每天的长度变长约每世纪2 ms。

(2) 数十年尺度上的日长波动,由地球液核运动和缓慢气候变迁所致。

(3) 固体潮和海潮引起的约 1 ms 的变化,有长(每年)和短(每月的和更短的)周期部分。

(4) 季节性效应,解释为大气激发所致,也有水及冰平衡变化的贡献。

(5) 更多不规则振荡,由不同的源造成,如地球质量位移(地震),太阳活动和大气事件,例如厄尔尼诺现象。

极移对观测的影响与观测位置有关,而 LOD 变化对所有点的影响是相同的。极移坐标和 LOD,以及 ω,由 IERS 以地球定向参数 EOP 提供,精度分别为 ±0.000 3″和 ±0.02 ms 或更好。

地心(地球参考系的原点)位置相对于监测台站随时间有微小变化。从卫星轨道的分析中已经发现了年度变化和半年变化,幅度为每年几个毫米,这些变化主要由大气及海洋的质量重新分配引起,以及由大陆水变化引起。通过 ITRF 站坐标给出的地心精度为几个毫米。

2.4.3 国际地球参考框架

IERS 通过一组全球分布的空间大地测量观测站点来实现 ITRS。观测站的地心直角坐标和速率构成了 ITRF。纳入 ITRF 的测站开展连续观测或者按一定的时间间隔开展观测。观测是在 12 个较大的构造板块上进行的,这样可以得到跟板块构造有关的站速率。

ITRF 的年度实现由 IERS 发布,ITRF97 包括约 320 个台站上的 550 多个测站的地心坐标 (X, Y, Z) 以及相应的台站速率。结果的精度跟观测技术有关,且 VLBI、SLR 和 GPS 观测的精度最高。顾及了几种时变效应,包括由固体潮、海洋和大气负载以及冰后回弹引起的测站位移。ITRF 满足相对于板块构造模型 NNR-NUVEL1A 无绝对旋转条件,也完全不允许有垂直运动。ITRF 定向相对于 IERS 参考极和参考子午圈给定。地球表面的实际(历元 t)点位位置矢量 r 根据它在参考历元(t_0)的坐标由下式导出

$$r(t) = r_0 + \dot{r}_0(t - t_0) \qquad (2.7)$$

式中,r_0 和 $\dot{r}_0$ 分别为历元 t_0 的位置和速率。

2.5 与重力场有关的参考系

大多数在地球表面或接近地球表面的大地和天文观测,都是通过沿本地铅垂线方向参照于地球重力场的,因此,须引入本地与重力场有关的参考系来对观测量建模。相对于全球参考系的本地系统定向由天文经纬度给定,这些定向参数用于实现从本地系统到全球系统的转换,以及反向转换。

2.5.1 本地铅垂线的方向

相对于全球地心系的铅垂线(本地铅垂线)方向由两个角表示(见图2.4)。天文(地理)纬度 φ 为赤道面和过点 P 的本地铅垂线之间的夹角,在子午面内量算,从赤道向北量为正,向南量为负。在赤道面内量算,格林尼治子午面和过点 P 的子午面之间的夹角称为天文(地理)经度 λ,向东量为正。再加上点 P 处的重力位 W,就确定了点 P 在水准面系统中的位置,水准面可以表示为 $W = $ 常数。本地天文子午面是由点 P 的本地铅垂线和平行于自转轴的直线所张成的。

引入定义在重力场中的非线性"自然"坐标系 φ, λ, W。天文纬度 φ 和天文经度 λ 描述了点 P 的铅垂线方向,重力位 W 确定 P 点位于水准面 $W = $ 常数上。因此,P 点由非正交的坐标面 $\varphi = $ 常数,$\lambda = $ 常数和 $W = $ 常数相交确定。坐标线(空间曲线)分别称为天文子午线($\lambda, W = $ 常数),天文纬线($\varphi, W = $ 常数)和同天顶线($\varphi, \lambda = $ 常数)。

自然坐标可由观测确定。天文定位可得纬度和经度,尽管 W 不能直接测定,但可由水准测量和重力观测确定相对于某选定水准面(例如大地水准面)的位差。

2.5.2 本地天文系统

天文和大地观测与观测点处的铅垂线方向相联,因而与地球重力场相联,有一种例外的情形是距离测量,跟参考系没有关系。因此,这些观测量构成了与本地重力场有关的系统:本地天文系统。其原点位于测站点 P,z 轴与本地铅垂线重合且指向天顶,x 轴(北方向)和 y 轴(东方向)张成水平面,与水准面 $W = W_P$ 相切,该坐标系为左手系。

2.5.3 本地天文系统和地心体系之间的转换

铅垂线相对于全球地心体系可用"定向"参数天文经纬度 λ 和 φ 表示(见图2.5)。把全球系平移到本地系之后,应用反射矩阵把本地系变换为右手系。

$$S_2 = \begin{pmatrix} 1 & 0 & 0 \\ 0 & -1 & 0 \\ 0 & 0 & 1 \end{pmatrix}$$

然后绕(新) y 轴将本地系旋转 $90° - \varphi$，绕 z 轴旋转 $180° - \lambda$，对应旋转矩阵为

$$R_2(90° - \varphi) = \begin{pmatrix} \sin\varphi & 0 & -\cos\varphi \\ 0 & 1 & 0 \\ \cos\varphi & 0 & \sin\varphi \end{pmatrix}$$

$$R_3(180° - \lambda) = \begin{pmatrix} -\cos\lambda & \sin\lambda & 0 \\ -\sin\lambda & -\cos\lambda & 0 \\ 0 & 0 & 1 \end{pmatrix}$$

由此可得 P_i 和 P 在地心体系下的坐标差为

$$\Delta X = Ax \tag{2.8}$$

x 由下式给出

$$x = \begin{pmatrix} x \\ y \\ z \end{pmatrix} = s \begin{pmatrix} \cos A \sin z \\ \sin A \sin z \\ \cos z \end{pmatrix}$$

且

$$\Delta X = \begin{pmatrix} \Delta X \\ \Delta Y \\ \Delta Z \end{pmatrix}$$

变换矩阵为

$$A = R_3(180° - \lambda) R_2(90° - \varphi) S_2 = \begin{pmatrix} -\sin\varphi\cos\lambda & -\sin\lambda & \cos\varphi\cos\lambda \\ -\sin\varphi\sin\lambda & \cos\lambda & \cos\varphi\sin\lambda \\ \cos\varphi & 0 & \sin\varphi \end{pmatrix} \tag{2.9}$$

式(2.8)的逆阵容易求得，顾及 A 为正交阵

$$A^{-1} = A^T$$

可得

$$x = A^{-1} \Delta X \tag{2.10}$$

且

$$A^{-1} = \begin{pmatrix} -\sin\varphi\cos\lambda & -\sin\varphi\sin\lambda & \cos\varphi \\ -\sin\lambda & \cos\lambda & 0 \\ \cos\varphi\cos\lambda & \cos\varphi\sin\lambda & \sin\varphi \end{pmatrix} \tag{2.11}$$

从式(2.8)至式(2.11)为计算三维参考框架下本地大地测量观测量的基本公式。

3 大地测量中的观测技术

3.1 地面大地测量观测

地面大地测量观测是在地球表面开展的，直接观测点与点之间的几何量。大多数观测量通过本地天文系统的定向，参照于地球重力场。水平角、天顶距和距离测量可用于相对定位，

一般使用组合式测量仪器(全站仪)观测;精密高差由水准测量测定。

由于卫星测量技术的高精度和经济性,地面大地测量观测主要用于卫星定位点位的加密,或者是用于卫星方法失效或需要地面测量支持的地区(地下和水下定位,森林,城市地区,工程测量,本地地球动力学工程等)。

3.1.1 水平角测量

三角测量

在导航、测量和土木工程中,用于精确确定船只或飞行器的位置,以及道路、隧道或其他在建结构方位的一种技术。三角测量依据平面三角学定理,即若三角形的一条边和两个角已知,其他的两条边和一个角度都可以容易地算出。测量出所选定三角形中一条边的边长,作为基线,再用测量仪器,如经纬仪,测出两个邻角,就可确定整个三角形。通过构造一系列的这种三角形,每个三角形至少与一个其他三角形相联,则可确定未观测边角的值。在很早以前,古埃及、古希腊和其他民族都用到了三角测量,所用的仪器是粗糙的瞄准器,后来改进成早期经纬仪。这些在公元1世纪希罗(Heron)的著作中就有描述(希罗,公元1世纪,希腊数学家和发明家,通称"亚历山大的希罗")。

经纬仪

基本测量仪器,起源未知,可追溯到16世纪英国数学家Leonard Digges,用来测量水平角和垂直角。现代经纬仪包含一个可水平和垂直旋转地望远镜,通过水准器来进行整平,望远镜中的十字丝用于精确照准目标。望远镜经过精密调整后,读取对应的水平和垂直刻度读数。

在道路修建、隧道准直和其他土木工程建设中,将经纬仪安置在可调节的三脚架上,在野外获取三角形的精密角度观测量。子午仪是一种特殊的经纬仪,其望远镜装置成能够完全倒转的形式。摄影经纬仪,组合了照相机和经纬仪,架设在同一个三脚架上,用于制图或其他用途的地面摄影测量。

3.1.2 距离测量

三边测量

一种测量三角形边长的测量方法,通常用电磁手段,根据边长信息可以计算角度。通过构建一系列相邻的三角形,可以获取其他不可量测的边长和角度。以前,相比三角测量三边测量很少用到,原因在于计算较困难。三角测量是已知三角形的一边和两个内角确定另外两边和一角的方法。电磁测距设备的发展使得三边测量成为一种常见的、首选的方法。三边测量的外业测量程序与三角测量类似,区别在于,仅仅观测边长,而角度是由计算得到。

电磁波测距

过去在大范围地区或山区开展距离测量是不切实际的,直到20世纪中期电磁波测距技术(EDM)的发明。这项技术通过在被测距离上对电磁辐射传播计时,使得距离测量像角度测量一样准确和容易;微波可以穿透大气雾霾,常用于长距离测量,光波和红外辐射用于短距离测量。用于EDM的设备中,辐射信号用的要么是光束(用激光器或电灯产生),要么是超高频无线电束。光束需要畅通的视线,无线电束能穿透雾、烟、大雨、灰尘、沙暴和枝叶。在主动站上,这两种类型都需要收发装置,在被动站上,光束测量需要一组反射棱镜;高频无线电束测量需要一台同主动站上收发装置相同的转发器(需要一名操作员)。反射棱镜的形状是正方体一角的内部,在合理的限度内,能够将任何方向入射的光线反射回光源方向,此时转发器必须面向收发装置。

这两种设备类型,都是通过对光束或无线电束的往返时间长度测量来测距,传播时间又是通过叠加在载波上的调制信号的相位变化来确定,根据电子电路学确定相位变化,并转换为时间。因为仅用一个频率可能会产生模糊度,用多个调制频率可以消除模糊度。

电磁波测距仪

电磁波测距设备有光电测距仪和微波测距仪,这些设备可测量斜距,斜距可通过大地高以足够的精度归算到椭球面。

3.1.3 三角高程测量

在山区,为了快速测量高差,且对精度要求不高时,可采用三角高程测量,即用经纬仪测量垂直角,通过测量距离或用三角测量方法计算距离。当大多数控制点在山顶时,在三角测量或导线手段布设的主要控制网区域,用该方法获取高程特别有用。为了提高精度,采用同时对向观测以消除空气折射,这种做法最好在午间实施,因为这时大气已混合均匀。

3.1.4 水准测量

几个世纪以来,测量员用水准仪进行水准测量,水准仪包含一个装有十字丝的水平望远镜,望远镜可以绕三脚架上的垂直轴旋转,还有一个十分精密的水准器,通过使气泡精确居中来整平水准仪,通过望远镜观测带有刻度的竖直标尺上的读数。如果将标尺依次的放置在地面点上,且望远镜是真正水平的,则十字丝的读数之差就等于点位的高程之差。在一定路线或道路上交替移动水准仪和标尺,并重复这一程序,就能精确测出较长水平距离上的高差。

3.2 卫星测量

卫星大地测量利用人造卫星和月球作为地外目标或传感器。自20世纪60年代到80年代引入并采用的经典卫星观测技术,在建立大范围大地控制网和重力场确定方面表现出其效能。现在,GPS适用于所有尺度的三维定位,而卫星激光测距主要服务于全球参考网。卫星测高通过监测海平面对重力场建模做出贡献;而通过未来的卫-卫跟踪计划和重力梯度测量计划,有望实现高分辨率全球重力场的恢复。

3.2.1 GPS

GPS由美国国防部设计、建造、管理和维护。第一颗GPS卫星于1978年发射,系统在20世纪90年代中期正式运行。GPS星座包括24颗卫星,分布在6个轨道面上,每个轨道有4颗卫星。各轨道平面的升交点间隔40°,倾角为55°,每颗GPS卫星在近圆轨道上运行,轨道长半径为26 578 km,周期约为12 h。卫星连续进行自定向,保证其太阳电池板保持指向太阳,天线指向地球。每颗卫星携带四台原子钟,卫星大小如同一台小汽车,重约1 000 kg。卫星钟的长期频率稳定性优于10^{-13}/天,卫星上的原子钟产生L波段10.23 MHz的基频。

GPS卫星由5个基准站监测,主控站位于科罗拉多州的科罗拉多斯普利斯,其他四个分别位于阿森松岛(大西洋)、迪戈加西亚(印度洋)、夸贾林岛和夏威夷(两个都在太平洋)。所有基准站都安装有精密铯钟和接收机,用于确定广播星历,并对卫星钟建模。将星历和钟差改正发送给卫星,卫星再把更新信息放到信号中发送给 GPS 接收机。

每颗 GPS 卫星在三个频率上发送数据:L1(1 575.42 MHz)、L2(1 227.60 MHz)、L5(1 176.45 MHz)。L1、L2、L5 载波频率分别由基频乘以 154、120 和 115 得到。伪随机(PRN)码和卫星星历、电离层模型以及卫星钟改正一起被调制在 L1、L2、L5 载波频率上。测得信号从卫星到接收机的传播时间,用来计算伪距。粗测距(C/A)码,有时也称为标准定位服务(SPS),是一种调制在 L1 载波上的伪随机码。精测距(P)码,有时也称为精密定位服务(PPS),调制在 L1、L2、L5 载波上,可以用来消除电离层效应。

一般把 GPS 当作一种测距系统,从已知位置的空间卫星到陆地、海洋、空中和太空中未知点的测距系统。GPS 卫星轨道可由广播星历或 IGS 获得。IGS 轨道是经后处理或准实时处理后的精密星历。所有 GPS 接收机都有植入其计算机的星历,星历可以提供每颗卫星在任何时刻的位置。星历是包含所有卫星轨道和钟差信息的数据文件,由 GPS 卫星发送给 GPS 接收机,帮助 GPS 接收机快速捕获人造卫星。GPS 接收机探测、解码、并处理从卫星接收到的信号,得到码、相位和多普勒观测数据,数据可以实时获取或者存储下来以供下载。接收机内置软件通常用于处理实时数据,采用的是单点定位方法,并为用户输出信息。由于接收机软件限制,精密定位和导航通常由带有功能更强大软件的外部计算机来实施。

GPS 的基本功能是告诉用户在哪里、如何运动以及什么时间等信息。由于 GPS 技术进入民用领域,其应用已变得几乎没有止境,理解 GPS 成为必需。

3.2.2 SLR

激光已被用来测量地球表面的距离,还用于计算从地面站到卫星和月球表面的距离。激光设备照准目标,然后由时钟在适当的时间激活,激光束通过目标上的特殊反射装置反射回来,回光通过光电方法探测到,测量出飞行时间然后转化为距离。激光发射器装置在接收反射激光束的望远镜或光学设备附近。

在卫星激光测距中,发射脉冲和从卫星返回脉冲之间的间隔能被非常准确地测量,然后转化为经过大气折射改正的测量距离。即便是卫星位于地球的阴影中,以及在白天,激光测距都可实施。

两个测站用激光测距同步测量一颗近地卫星,固定一个激光站的位置,可确定另一个激光站相对于该站的坐标,以及两站间的距离。自 1972 年,NASA 应用激光跟踪技术测量北美洲点与点之间的距离,同时,验证激光跟踪在测量圣安地列斯断裂两侧点与点之间地壳运动中的精度,并计划做跨断裂带基线的多年重复测量。同步激光跟踪在东海岸一个测站和百慕大站之间也获得了成功,使得确定百慕大站的相对位置(NAD 基准)以及两站间的基线成为可能。

史密森天体物理观测台(SAO)和美国国防部(DoD)还将激光测距数据应用于建立 WGS 坐标系,NASA 在建立重力场模型时也用到了卫星激光数据。卫星激光测距数据还应用于极移和地球自转的研究。

3.2.3 卫星测高

卫星测高基于星载雷达测高仪,测高仪向地面垂直的方向发射脉冲,海洋面垂直地反射脉冲,测量传播时间就可得到卫星相对于瞬时海面的高程(图 3.1)。

$$a = \frac{c}{2}\Delta t \tag{3.1}$$

球近似时,观测方程可表示为

$$a = r_s - r_P - (N + \text{SST}) \tag{3.2}$$

式中,r_s 和 r_P 为从地心到卫星和到椭球面上星下点 P 的距离;N 为大地水准面高,SST 为海面地形高。卫星跟踪给出 r_s,定位给出 r_P,因而,卫星测高可以求定大地水准面以及海面地形信息。

雷达测高仪的工作频率为 14 GHz,以短(几纳秒)脉冲和高脉冲频率(例如 100 脉冲/秒)方式测距。射束发散效应和有限脉冲长度使得测量的是基于一定圆"足迹"(几 km 范围)的"平均"表面;因此海洋(浪)的短波特征被平滑。观测量必须加入仪器改正(通过实验室和检验区域校正),大气影响改正,以及海潮改正(潮汐模型)。顾及海洋学和气象学类型的时变,以及水量平衡,可得到"拟稳态"海面地形,再精确确定轨道位置,则可在地心坐标系下建立该海面地形的位置。在更长的卫星观测任务中,可以以近似一致的剖面对海平面进行多次观测。

Skylab 任务成功完成验证之后,GEOS 3 卫星完成了使用雷达测高技术进行第一次全球观测。它平均 2 s 得到一个观测量(足迹间隔为 15 km),精度为 ±0.5 m。在 Seasat 1 任务(1978)中,通过改用更短的脉冲波长,使测量精度提高到 ±0.1 m。

ERS-1 卫星和 TOPEX/POSEIDON 卫星的雷达测高仪观测精度优于 10 cm。

3.2.4　卫-卫跟踪,卫星重力梯度测量

从太空确定高分辨率重力场需要低轨卫星和高敏传感器,可由卫-卫跟踪和卫星重力梯度测量实现。

卫-卫跟踪(SST)采用微波系统测量两颗卫星之间的距离变化率。设计有高-低(一颗高轨卫星和一颗低轨卫星)和低-低(两颗低轨卫星在同一高度)两种形式。基本观测量包括距离变化率(径向速度)以及距离变化率的变化,后者是由引力和非引力"摄动"力引起的。通过对非引力摄动力的正确补偿之后可导出引力场参数(球谐系数)。为了得到 100 km 的重力场分辨率,卫星的轨道高度不得超过几百千米,卫星之间的相对速率必须以 $\pm(1 \sim 10)\ \mu\text{m/s}$ 的精度测定,且要保证由高轨卫星系统(GPS)和地面站的精确跟踪。

卫星重力梯度测量确定重力梯度张量(重力位的二阶导数)的各分量。在地球表面,约从 1900 年起已经实施了重力梯度测量,工具是利用在某方向对本地重力场变化敏感的传感器对(加速度计)。采用不同定向的传感器,可以确定重力梯度的不同分量。对于空载应用,由于重力场随高度衰减(在几百公里高度,梯度张量的非对角元素只有几个 10^{-9}s^{-2}),需要对二阶导数的高精度测量(为 10^{-11} 至 $10^{-13}\ \text{s}^{-2}$ 量级),这需要用常规或超导电子学才能实现。对传感器对的姿态控制和加速度计的漂移提出了很高的要求。加速度计对的输出求差之后可消除非引力摄动力。

3.3　重力测量

"绝对"重力测量是指直接参照于时间和长度基准,而"相对"测量是一种反作用力来确定重力差。为了使本地或区域重力网参照一个统一基准,需要建立全球重力参考系。移动平台

上的重力测量对于施测困难地区是很有用的。

在国际单位制中,重力的单位为 ms^{-2}。在大地测量学和地球物理学中,单位 mGal = 10^{-5} ms^{-2} 和 μGal = 10^{-8} ms^{-2} = 10 nms^{-2} 仍广泛应用,它起源于之前的"厘米 – 克 – 秒制(cgs)"中的单位 Gal(以伽利略之名命名)。

在地球表面,重力加速度 g 为

$$g \approx 980 \text{ cm/s}^2 \equiv 980 \text{ Gal}$$

测量和观测静态重力需要约 1 mGal(相对于重力加速度精度为 10^{-6})的精度。为研究重力变化,需要 1 μGal 的测量精度,例如,若升高 3 mm,重力减小(因为离地球质心更远)约 1 μGal。因此,1 μGal 精度相应于约 3 mm 的垂直位移。大多数长期垂直地壳运动,其量级为每年几个毫米,故在这一精度水平的重力测量可用来估算垂直运动。

有两种类型的重力仪:

绝对重力仪,重力 g 可通过测量长度和时间直接确定。

相对重力仪,重力 g 跟弹簧常数有关,该常数不太容易测定。相对重力仪只能给出两点之间或两个时刻之间的相对重力差。

重力参考系

大地测量学和地球物理学中所需的重力值,必须参照于一个全球参考系。重力参考系定义为许多精确测定重力的控制点的重力值。

1909 ~ 1971 年,波茨坦重力系统充当国际参考系,它的建立基于波茨坦大地测量研究所施测的可倒摆测量(1898—1904 年)。更多的现代绝对重力测量证实波茨坦重力值大了 14 mGal。因此,1950 ~ 1970 年,通过国际联合建立了新的国际重力系统。

1971 年在莫斯科 IUGG 代表大会上,引入 1971 国际重力基准网(IGSN71)作为新的参考系,该网包含约 1854 个点(约 500 个基本站),重力值由 10 个新的绝对重力观测量和约 25 000 个相对重力观测量(包含约 1 200 个相对摆观测),整体不确定性小于 ±0.1 mGal。在几条南北走向(重力差较大)的重力测量检定基线(欧洲-非洲、美洲、西太平洋)上,进行了密集地、特别高精度地观测。参照于其他系统的区域重力网应当联结到 IGSN71,转换参数(一般是平移和尺度参数)通过公共点推求,在公共点上有两个系统的数据。

IAG 建议建立国际绝对重力基准站网(IAGBN),包含 36 个全球分布的测站,该网的主要目的是在全球尺度上监测重力时变,同时作为区域重力控制。该网自 1987 年起,利用移动式绝对重力仪建设。

3.4 天文测量

经典大地天文学涉及天文纬度、天文经度和天文方位角的确定,采用对恒星的地基光学测量手段和测时。

大地天文学是基于球面天文学的,自从发展了高效的卫星定位和重力测量手段之后,其重要性已经降低,现在主要限于局部重力场应用(铅垂线方向)和方位角测定;另外,VLBI 技术观测从河外射电源发出的无线电波,广泛应用于测得基本地面站之间的基线向量和确定地球自转参数。

天文纬度、天文经度和天文方位角的测定基于大地天文学给出的关系,其中恒星位置由星表给出。这里,对大地天文学中发展的方法就不做介绍,如赫瑞鲍-太尔各特法(Horrebow-Talcott method)。

对河外射电源的观测,如类星体,可提供大地测量信息确定距离非常远的两个射电望远镜天线之间的矢量,矢量由长度和方向组成。为了完成这一任务,必须要非常精确地测量从某一给定射电源发出的特定波前面的到达时间,到达时间由两个观测站的天线记录。在 VLBI 中,将记录下的信号电子叠加,生成合成扰动或干涉图样,这种现象称为干涉。相对相位延迟的理论公式表明,它是射电源方向、天线位置、两测站的相对钟差、时刻、所采用的大气模型、地球潮汐参数、观测所用的射电频率等因素的函数,还必须正确考虑地球自转。VLBI 技术的两个主要限制因素是时钟稳定性和大气变化。VIBI 的主要目标是,将跨大洲基线的不确定性降低到厘米级。

VLBI 技术得到的基线已用于为 1972 年建立的 DoD WGS 提供尺度信息。精确到厘米级的基线,可以作为未来世界坐标系的比较标准。VLBI 的其他应用包括:确定极移,地球自转变化,以及监测构成地球地壳的主要板块的运动。

4 大地控制网

大地控制网和重力控制网包括埋石控制点,在各种尺度上为定位和重力场测定提供参考框架。全球控制网可以实现按国际协议定义的参考系;区域控制网构成了国家或跨国家(洲际)大地测量和重力测量的重要基础,是各种地理信息系统和各种系列地图的基础;局域控制网是为工程测量、勘探性项目和地球动力学研究而特别建立的。

下面主要讨论区域控制网,现在区域控制网日益融入全球参考框架,全球参考框架是指为定位、导航和重力测量而建立的全球网。按照经典定位和测高的处理方法,平面控制网和高程控制网是分开来建立的,这些控制网仍然是国家大地测量系统的基础。空间大地测量方法可以在地心参考系下建立三维控制网,现在正在取代经典大地控制网。重力控制网满足了大地测量学和地球物理学的需要,参考系由全球控制网或通过绝对重力测量提供。

4.1 平面控制网

国家平面控制网是从 18 世纪到 20 世纪 60 年代之间建立的,控制网的设计、观测和计算是随着可用技术而不断转变的。计算是在与测量地区吻合的参考椭球面上开展的。自 20 世纪 60 年代起,空间大地测量方法已使经典控制网相对于全球参考系定位。

4.1.1 设计、埋石、观测

平面控制网由三角(三角测量)点实现,原则上三角点应该在国家范围内均匀分布。三角点可区分为不同等级,从一等点(站点间距从 30~60 km)到二等点(约 10 km),再到四等甚至是五等点(降到 1 至 2 km)。一等点间的最大间距由地面测量手段决定,需要控制网点间相互通视。因此,一等和一些二等点建立在丘陵和高山的顶端;特别是在平原地区,要搭建观测塔(木质或钢结构,高度 30 m 或更高)。站点通过地下和地面埋石(石板、石头或混凝土标石,坚实基岩上的螺栓)作永久性标志。为了帮助修复和核实中心标志,还要设立偏心标志。

经典一等和二等平面控制网用三角测量、三边测量和导线测量等手段施测。

在三角测量中,用经纬仪测量由三角点构成的三角形的所有内角,仪器架设在观测墩或测塔上;在较远距离上,通过目标点上的照明信号帮助照准。测量数个测回(即望远镜的盘左盘右)的方向(连续观测所有目标点)或角度(分别测量构成一个角的两个方向),各测回分布在经纬仪的水平度盘上。三角网的尺度通过至少一条三角形边长确定,边长可以通过短基线经基线扩展得到,也可以用距离测量仪器直接测得。天文观测提供控制网的定向,水平定向需要天文方位角按照拉普拉斯方程实现,该点称为拉普拉斯点。在扩展网中,基线和拉普拉斯点一般设在间隔几百千米的距离上,目的是控制测量误差相对于尺度和定向(横向折射影响)随控制网传播。

三边测量采用电磁波测距仪,测量网中所有三角形边长,包括对角线。至少需要一个拉普拉斯方位角用于网的定向。与角度测量相比,电磁波测距对测塔的稳定性要求要低一些,再者,激光和微波的应用使得距离测量受气象条件影响较小。

导线测量联合了角度和距离测量,导线点沿纵向展开。同样,为了定向导线,需要拉普拉斯方位角。导线测量是建立平面控制网的一种高效、灵活的方法,对于点建在制高点没有太多要求。它主要用于对较高等级控制网的加密,由已有控制点的坐标提供定向。

平面控制网也可以组合三角测量、三边测量和导线测量进行布设。

4.1.2 内业计算和定向

经典一等平面控制网和一些二等网是在参考椭球上大地坐标系下计算的,低等网经椭球正形投影到平面后,主要在平面直角坐标系中计算。

野外测得的水平角、方向和斜距首先要归算到椭球上,在早期测量中,没有考虑与重力场有关的归算(垂线偏差,大地水准面高)。有了冗余观测,如三角形闭合冗余,三边测量中的四边形对角线冗余,多余基线和拉普拉斯方位角冗余等,网平差可以按照条件平差法实施,也可以按照参数平差法实施。对于条件平差法,首先要对控制网的几何条件进行平差,然后在椭球面上按照大地主题解算的正解,从大地原点开始计算坐标;而对于参数平差,要从大地主题解算的反解导出椭球"观测方程"。经典"展开法"的不足在于,忽略垂线偏差,距离到椭球面的归算不充分,特别是大规模控制网的分步计算,当把一个新的子网联系到已有控制网时,会有联结限制,会导致控制网的某种扭曲,产生区域性的尺度误差($\pm 10^{-5}$)和定向误差($\pm$几个弧秒)。相对于大地原点的相对坐标误差,已经从约 100 km 处的几个分米,增大到了几百千米处的 1 m 左右,到广大扩展网的边缘可达 10 m 或更多。

4.2 高程控制网

高程控制网是与平面控制网分开来建立的,因为高程必须要参照于重力场,而不是水平定位所需要的椭球系统。

高程控制网通过几何水准来建立,有时也采用静水位水准(基于连通器原理)。根据水准测量规程以及所能达到的精度,国家大地测量将水准测量分为几个等级。按照精密水准规定,一等水准构成闭合环状(环线周长为几百千米),水准环由水准路线构成,水准路线连接网的各个节点,水准路线又由连接相邻水准点的测段构成。一等水准网由二等至四等水准测量加密。

水准路线通常沿主要公路、铁路和河流布设,水准点由建筑物、基岩、或混凝土桩上的螺栓构成。在冲积地区,要建立长的管道;在地质稳定地区,为了保持控制网随时间的稳定性,要建立地下标石。一等水准网可得到约每千米 1 mm 的精度。每隔大约十年,应对一等网重新观测,因为本地性和区域性高程变化可达 1 mm/a 或更多,特别是存在地壳运动的地区更是如此。

在进行水准网平差之前,要考虑表面重力的影响,必须把原始观测高差转换成重力位差,或者是正高或正常高高差。然后,应用环闭合差条件,按条件平差法或者采用更可取的参数平差法进行网平差。

高程基准(零高程面)定义为由验潮记录确定的平均海平面(MSL)。不同国家的高程系统之间可能相差数分米至 1 m,或者更多。由于海面地形的影响,高程基准也不同于作为全球参考面的大地水准面。如果高程基准约束到多个验潮站的 MSL 则会导致网的扭曲。

4.3 三维控制网

空间大地测量技术能够在本地、区域和全球尺度上,在全球地心体系下以厘米级精度传递三维坐标。全球控制网实现了地球参考系,并提供建立大洲级控制网的参考点,其中 GPS 具有重要的地位。进一步的控制网加密几乎毫无例外的采用 GPS 技术,而且把已有经典控制网集成起来。

ITRF 是全球三维定位的基础,由一组全球空间大地测量台站的地心直角坐标及水平速率定义,给定某参考历元,精度分别为每年约 ±1 cm 和 ±1~3 mm。ITRF 结果基于采用不同空间技术的全球网所采集的观测量,在 IERS 框架下联合处理得到。

自 20 世纪 90 年代起,采用 GPS 手段建立国家大地控制网,现在取代了经典水平(高程也有可能)控制网。尽管现在仍在讨论建立和维持这些三维控制网的策略,并且各个国家的情况不尽相同,但还是可以清楚地总结出以下趋势:

(1) 建立站间距在 10~50 km 的大尺度三维网;
(2) 间距在 50~100 km 的连续运行 GPS 站建设;
(3) 最终利用新的 GPS 控制,将已有平面控制网进行转换。

4.4 重力控制网

重力网提供在全球、区域和本地尺度上进行重力测量的框架,包括用绝对和相对重力测量方法测得重力的重力点。在全球尺度上,重力基准由 IGSN71 实现,但现在可以使用绝对重力仪独立建立基准。全球绝对重力基准站网的建立主要是为了研究重力随时间的长期变化。

国家重力测量以基本网或基准网为基础,大多数情况下由低等网加密。重力基准网点应在区域内均匀分布,对于范围较大的国家,点间距一般为几百千米。台站所处的地质、水文、微震环境应当稳定,且应位于永久性场所(天文台等)。偏心点用于保护中心点,以及用于控制本地高程和质量变化。后续的加密网点可以与水平或高程控制点并址。网点的水平位置和高程应分别以米级和毫米至厘米级精度测定。

5 地球重力场

在大地测量学中,外部重力场具有重要的作用,因为地球形状是在重力作用下演化的,且大多数大地测量观测量是参照于重力场的。因而,建立观测量的模型需要理解重力场。此外,对地球重力场的分析会得到关于地球内部结构的信息,这样,大地测量学对地球物理学也有贡献。

5.1 重力场基础

随地球一同自转的物体,受到地球质量的引力和地球自转产生的离心力作用,其合力称为重力。对于人造卫星,注意到卫星不随地球自转,因此只有引力作用到卫星。

5.1.1 引力和引力位

根据牛顿万有引力定律,两个点质量 m_1 和 m_2 相互吸引,万有引力(吸引力)为

$$F = -G\frac{m_1 m_2}{r^2}\frac{r}{r} \tag{5.1}$$

式中,G 为万有引力常数;r 为点质量之间的距离。矢量 F 和 r 方向相反。

如果使用标量"位",而不是矢量"加速度"的话,则重力加速度、重力场以及有关的计算都会得到简化。

对于点质量 m,其位为

$$V = \frac{Gm}{r} \tag{5.2}$$

对于地球,有

$$V = G\iiint_{\text{earth}} \frac{\mathrm{d}m}{r} = G\iiint_{\text{earth}} \frac{\rho}{r}\mathrm{d}v \tag{5.3}$$

点 P 的位表示为把单位质量从无穷远处($V = 0$)移至点 P 引力所做的功。位的单位为 m^2s^{-2}。

5.1.2 离心力和离心力位

离心力作用于地球表面的质体,是由地球绕轴自转产生的。假定绕自转轴的自转角速度 ω 恒定,且假定自转轴相对地球固定,则离心加速度

$$z = (\omega \times r) \times \omega = \omega^2 p \tag{5.4}$$

表示的是作用在单位质量上的、与自转轴垂直且指向外侧的离心力(图 5.1)。

由

$$z = \operatorname{grad} Z \tag{5.5}$$

引入离心力位为

$$Z = \frac{\omega^2}{2}p^2 \tag{5.6}$$

5.1.3 重力加速度和重力位

重力加速度,即重力 g,是万有引力 F 和离心力 z 的合力

$$g = F + z \tag{5.7}$$

重力 g 的方向是指铅垂线(垂线)的方向;其大小称为重力强度(常称为重力)。由式(5.3)和式(5.6),地球重力位为

$$W = V + Z = G\iiint_{\text{earth}} \frac{\rho}{r}\mathrm{d}v + \frac{\omega^2}{2}p^2 \tag{5.8}$$

重力位与重力的关系为

$$g = \mathrm{grad}W \tag{5.9}$$

5.2 水准面和铅垂线

重力位为常数的面

$$W = W(r) = \text{const.} \tag{5.10}$$

叫做重力等位面或水准面(也称等位势面)。对于无穷小位移 $\mathrm{d}s$,两无限接近的水准面之间的位差为

$$\mathrm{d}W = g \cdot \mathrm{d}s = g\mathrm{d}s\cos(g,\mathrm{d}s) \tag{5.11}$$

该式表示重力位对某方向的导数等于重力在该方向的分力。由于在(5.11)中,只有 $\mathrm{d}s$ 沿铅垂线方向的投影得以保留,故 $\mathrm{d}W$ 与路径无关。因而,在水准面 $W = \text{const.}$ 上的位移不需做功,水准面是等势能面。

若 $\mathrm{d}s$ 取为沿水准面 $W = W_P$ 方向,由于 $\mathrm{d}W = 0$,则有 $\cos(g,\mathrm{d}s) = \cos 90° = 0$,重力与水准面 $W = W_P$ 正交,水准面与铅垂线的交角为直角,铅垂线的切线称为铅垂线方向。若 $\mathrm{d}s$ 取为沿水准面外法线 n 方向,由于 $\cos(g,n) = \cos 180° = -1$,则得如下重要关系

$$\mathrm{d}W = -g\mathrm{d}n \tag{5.12}$$

该式建立了相邻水准面位差(物理量)和高差(几何量)之间的关系。根据该关系,联合重力测量和几何水准测得的高差,就能得到重力位差。

5.3 引力位的球谐展开

由于地球的密度函数 $\rho = \rho(r')$ 未知,引力位 $V = V(r)$ 不能按照牛顿万有引力定律计算。但是,将 V 展开为外部空间的收敛级数,作为拉普拉斯微分方程的特解,是可行的。这种解相当于将地球重力场做谱分解。级数展开的系数给出了相应谱分量的幅度。V 的任何可观测函数,都可以用于数值求解级数系数,因此,可以将重力场表示成全球分析解。

引力位展开成球谐级数形式可表示为

$$V = \frac{GM}{r}\left[1 + \sum_{l=1}^{\infty}\sum_{m=0}^{l}\left(\frac{a}{r}\right)^l(C_{lm}\cos m\lambda + S_{lm}\sin m\lambda)P_{lm}(\cos\vartheta)\right] \tag{5.13}$$

球谐系数为

$$\left.\begin{array}{r}C_{l0} = C_l = \dfrac{1}{M}\iiint\limits_{\text{earth}}\left(\dfrac{r'}{a}\right)^l P_l(\cos\vartheta')\,\mathrm{d}m \\[2mm] \left\{\begin{array}{c}C_{lm}\\ S_{lm}\end{array}\right\} = \dfrac{2}{M}\times\dfrac{(l-m)!}{(l+m)!}\iiint\limits_{\text{earth}}\left(\dfrac{r'}{a}\right)^l P_{lm}(\cos\vartheta')\left\{\begin{array}{c}\cos m\lambda'\\ \sin m\lambda'\end{array}\right\}\mathrm{d}m \qquad m\neq 0\end{array}\right\} \quad (5.14)$$

5.4 大地水准面

在大地测量学、海洋学和固体地球物理学中,大地水准面具有根本重要性。在大地测量学和海洋学中,大地水准面作为描述大洲和海面地形的高程参考面;地球物理学利用大地水准面作为地球重力场的表示,揭示地球深层质量分布。

5.4.1 定义

地球重力场中的等位面,与不受干扰的平均海水面重合,并向大陆内部延伸。重力方向在每一点上与大地水准面正交。大地水准面是天文测量和大地水准测量的参考面。

* * * * * * * * * * *

假设海水是可以自由移动的、均质的物质,其运动仅受地球重力的支配,在达到均衡状态时,这种理想海洋的表面呈现为重力场中的水准面,再将它延伸至大陆内部(例如,通过连通器系统),这一水准面就定义为大地水准面。其方程由下式给出:

$$W = W(r) = W_0 \tag{5.15}$$

可以看出,大地水准面是一个闭合、连续的水准面,且部分延伸到地球固体部分内部。大地水准面弯曲在密度剧烈变化处表现出不连续性。因此大地水准面不是一个解析面,因而将它排除作为定位的参考面,然而,它非常适合作为位差或高差的参考面,位差或高差可由几何水准结合重力测量的方式得到。

5.4.2 平均海平面

所有潮汐周期的海面平均高度,通常用最少 19 年时间内每小时高度读数的平均来确定,亦称海面基准。

* * * * * * * * * * *

为了建立大地水准面作为高程的参考面,要用验潮仪(记录潮汐的自动记潮仪)在较长的时间内记录海平面位置并取平均,这样得到的平均海平面(MSL)可作为大地水准面的近似。

由于验潮站通常并不能无扰动地与海水联系起来,验潮读数常常要受到系统影响。海平面随时间变化,若其变化是周期性的,则可通过对海平面记录取平均来在很大程度上消除其变化。

卫星测高得到了参考瞬时海平面的远海数据,海平面相对于大地水准面的高程表示海面地形。这里,应区别瞬时海面地形和近似静止海面地形,近似静止海面地形是将瞬时海面地形加入与时间有关的变化改正后得到的。

这些变化包括海洋潮汐,海洋潮汐会相当偏离理论值,这是由于海深不等以及大洲阻碍了海水运动而引起的。在深海,潮涨低于 1 m,然而,在沿海地区,潮涨会达到几米高。以年度为周期,且量值可达到 1 m 的波动包括:气象因素(大气压、风)引起的波动,海洋学因素(洋流,由温度、盐度和气压引起的海水密度不同)引起的波动,以及由水量平衡(由冰雪融化、降雨等引起的水量变化)引起的波动。目前,由冰后海面上升引起的长期变化大约是 1 mm/a。

尽管海平面观测的平均年度值内符精度可达 ± 1 cm,仍可能会出现年与年平均值之间偶然偏离 ± 10 cm 或更高的情况(与气象学有关的效应)。

即便综合考虑了所有周期的、非周期的以及长期的变化,这只有在特殊情况下才有可能,平均海平面也不能构成地球重力场的一个水准面,在大范围来看,其偏差可达 1 m 或更多。

5.5 高程系统

在大地测量学、地图制图学和海洋学中,大地水准面是高程和深度(大洲和海底地形,以及海面地形)的参考面。点 P 在由该点重力位 W 确定的特定水准面上(见图 5.2),相对于大地水准面的重力位 W_0,P 点的"高程"由到大地水准面的位差的负值给定,称为地球位数 C。由式(5.12),得

$$C = W_0 - W_P = -\int_{P_0}^{P} \mathrm{d}W = \int g \mathrm{d}n \tag{5.16}$$

积分与路径无关,因此,P_0 是大地水准面上的任意一点,C 可由 P_0 和 P 之间沿任意路线上的几何水准和重力测量确定。

地球位数是描述重力场中质量(例如,水质量)特性的一种理想度量,在某些应用中可作为"高程"应用,例如在水利工程和海洋学中,更一般的应用受限于位的单位:$m^2 s^{-2}$,它与采用"米"为单位的公制高程系统相矛盾。

为了与以米为单位的高程数值取得某种一致,地球位数也采用重力位单位(gpu) $10 \ m^2 s^{-2}$,或记为 kGal·m。由于 $g \approx 9.8 \ ms^{-2}$,C 的数值比相应的高程数值约小 2%。

地球位数除以某重力常数可得力高 H^{dyn}。通常采用水准椭球面在纬度 45°处计算得到的正常重力 γ_0^{45}

$$H^{\mathrm{dyn}} = \frac{C}{\gamma_0^{45}} \tag{5.17}$$

H^{dyn} = 常数的平面为等势面。因此,位于同一水准面上的点具有相同的力高。但无法对力高进行几何解释,为了将水准测量高差转化为力高高差,需要加入较大的改正。由于以上原因,力高在大地测量学中没有得到广泛应用,但在海洋学中有应用。

国家或大洲的高程系统,以及基于它们的地形数据(地形图,数字地面模型),采用正高或正常高。

正高 H 定义为从地面点沿铅垂线方向到大地水准面的距离。这一定义相当于通常理解的"海拔高"。将式(5.16)中的右边展开为 H,沿铅垂线从 P_0(正高 $H = 0$)积分至 P(正高为 H),得

$$H = \frac{C}{\bar{g}}, \quad \bar{g} = \frac{1}{H}\int_0^H g \mathrm{d}H \tag{5.18}$$

$\bar{g}$ 为沿铅垂线的平均重力,在计算平均重力时需要地球内部的重力值,可通过引入地形质量密度分布模型来实现,由于对这种密度分布认识不足,计算正高的精度取决于密度模型的精度。此外,正高相等的点轻微偏离水准面,这是由水准面的不平行性造成的。正高代表了地形质量的几何特征,这一点弥补了以上不足。几何水准只需加入较小的改正就可转化为正高高差。

为了避免地形质量分布的假设问题,引入了正常高 H^N,并且在许多国家得到应用。用沿轻微弯曲正常铅垂线的平均正常重力 $\bar{\gamma}$ 代替式(5.18)中的平均重力 $\bar{g}$,有

$$H^N = \frac{C}{\bar{\gamma}}, \quad \bar{\gamma} = \frac{1}{H^N} \int_0^{H^N} \gamma \, \mathrm{d}H^N \tag{5.19}$$

$\bar{\gamma}$ 在椭球地球模型的正常重力场中计算。正常高的参考面为似大地水准面,似大地水准面与大地水准面接近,但不是水准面,在低海拔地区与大地水准面的偏差为 mm 至 cm 级,在高原地区偏差可达 1 m,在海洋上,大地水准面与似大地水准面事实上是重合的。

国家高程系统的零高程面(高程基准),一般定义为一定时间间隔内验潮记录所确定的平均海平面。这些参考面仅仅是接近大地水准面,由于海面地形和局部异常影响,它们间的偏差最大可达 1 m。

6 参考椭球和大地坐标系

大地测量地球模型作为实际地球表面的参考,一方面应当与大地水准面吻合得很好,从而可将非线性大地测量问题线性化;另一方面,模型的数学形式应当简单,才可能用封闭公式计算。这种模型不仅作为大地测量学和地图学的应用标准,还可作为天文学和地球物理学的应用标准,应当能满足这些学科的必需。

6.1 旋转椭球

在 18 世纪,就已引入旋转椭球作为地球的几何形状。椭球的几何性质可以用椭球面坐标和曲率简单地进行描述。

6.1.1 几何参数

旋转椭球通过绕短轴旋转的子午椭圆得到,因而椭球形状可用长半轴 a 和短半轴 b 两个几何参数描述。通常,用更适合级数展开的众多小量中的一个来代替 b:(几何)扁率 f、偏心距 ε、第一和第二偏心率 e 和 e',分别为

$$f = \frac{a-b}{a}, \quad \varepsilon = \sqrt{a^2 - b^2}, \quad e = \frac{\varepsilon}{a}, \quad e' = \frac{\varepsilon}{b} \tag{6.1}$$

这些量之间存在如下关系

$$\frac{b}{a} = 1 - f = \sqrt{1-e^2} = \frac{1}{\sqrt{1+e'^2}} \tag{6.2}$$

6.1.2 大地和空间直角坐标

通常确定位置的一种更加方便的方法是用空间直角坐标。

空间直角坐标系的原点在地球中心，X 轴和 Y 轴位于赤道面内，Z 轴与地球平均自转轴（例如，Z 轴指向 CIO 方向）重合，X 轴通过零子午圈，也就是格林尼治平均子午圈（BIH 采用的零子午圈），三个坐标轴互相垂直，构成右手坐标系。

大地坐标 B（大地纬度），L（大地经度）和 h（大地高）与空间直角坐标 XYZ 的关系由下式给出

$$X = (N + h)\cos B\cos L$$
$$Y = (N + h)\cos B\sin L \tag{6.3}$$
$$Z = [N(1 - e^2) + h]\sin B$$

式中，N 为卯酉圈（东西方向）曲率半径

$$N = a/\sqrt{1 - e^2\sin^2 B} \tag{6.4}$$

空间直角坐标通常在卫星定位中作为计算坐标系。

反解问题可以用迭代方法或封闭公式求解。两种解法都用到了短轴到任意一点的距离 P，即

$$P = \sqrt{X^2 + Y^2} \tag{6.5}$$

或者，由式(6.3)，

$$P = (N + h)\cos B \tag{6.6}$$

由式(6.3)和式(6.6)，可得

$$\frac{Z}{P} = \tan B\left(1 - \frac{e^2 N}{N + h}\right) \tag{6.7}$$

这个方程是两种解算方法的出发点。迭代法通常根据上面的方程先解算 B 来初始化，令 $h = 0$，可得

$$B^{(0)} = \arctan\left(\frac{Z}{P}\frac{1}{1 - e^2}\right) \tag{6.8}$$

然后，第 K 次迭代要连续计算以下数值：根据 $N = a/\sqrt{1 - e^2\sin^2 B}$ 计算 $N^{(K)} = N(B^{(K-1)})$，根据式(6.6)计算 $h^{(K)} = h(B^{(K-1)}, N^{(K)})$，根据式(6.7)计算 $B^{(K)} = B(N^{(K)}, h^{(K)})$。重复迭代，直到满足下面的不等式时停止迭代

$$|h^{(K)} - h^{(K-1)}| < a\varepsilon \text{ and } |B^{(K)} - B^{(K-1)}| < \varepsilon \tag{6.9}$$

式中，ε 为某一先验选定的数值。一旦求定了 B 和 h，L 可以根据式(6.3)中前两个方程中的任一式来求定，或者根据

$$L = 2\arctan\frac{Y}{X + \sqrt{X^2 + Y^2}} \tag{6.10}$$

封闭形式解可由式(6.6)和式(6.3)得到

$$P\tan B - Z = e^2 N\sin B \tag{6.11}$$

在这个方程中，N 也是 B 的函数，B 是唯一的未知数，将 N 代入，则式(6.11)变为

$$P\tan B - Z = \frac{ae^2\sin B}{\sqrt{1 - e^2\sin^2 B}} \tag{6.12}$$

右式当中的分子和分母同除以 $\cos B$，然后将等式两边取平方，则得

$$P^2\tan^4 B - 2PZ\tan^3 B + \left(Z^2 + \frac{P^2 - a^2 e^4}{1 - e^2}\right)\tan^2 B - \frac{2PZ}{1 - e^2}\tan B + \frac{Z^2}{1 - e^2} = 0 \quad (6.13)$$

这是 $\tan B$ 的四次（双二次）方程，方程中所有系数已知。四次方程有标准解法，一旦求得 $\tan B$，N 和 h 可分别由 $N = a/\sqrt{1 - e^2\sin^2 B}$ 和式(6.6)计算，经度 L 可直接根据式(6.3)或式(6.10)求取，从而完成反算。Paul 证实，封闭形式解法比迭代方法要快大约 25%。

6.1.3 投影到参考椭球

下面讨论用自然坐标 φ, λ, H 来建立 P 点在大地坐标系中的位置。将 P 点沿铅垂线（稍稍弯曲）投影到大地水准面，正高就是 P 点与其在大地水准面上的投影点 P_0 之间的距离，沿铅垂线量取（见图 6.1）。尽管这种投影方式非常自然，但是大地水准面不适合直接在其上面进行计算，因而，再将点 P_0 沿椭球法线投影到参考椭球上，从而得到椭球面上一点 Q_0。用这种方式，地面点 P 及其在椭球面上的对应点 Q_0，就通过双次投影联系起来，即，通过两次连续的、相似的投影，得到正高 $H = PP_0$ 及其相应的大地水准面高 $N = P_0 Q_0$。这种双次投影称为毕兹特投影（Pizzetti's projection）。

更简单的方法是，把点 P 从地球自然表面直接沿椭球法线投影到椭球面上，得到点 Q，距离 $PQ = h$ 是相对于椭球面的几何高程。这样，地面点 P 就由 h 和椭球面上点 Q 的地理坐标 B, L 来确定，从而，由所谓的大地坐标 B, L, h 来代替自然坐标 φ, λ, H。这种投影称为赫尔默特投影（Helmept's projection）。

毕兹特投影和赫尔默特投影之间的实际差别很小。大地高 h 和 $H + N$ 的差别在十分之几个毫米以内，两种投影方式的大地坐标 B 和 L 之间的关系可由下式给出

$$\begin{aligned} B_{\text{Helmert}} &= B_{\text{Pizzetti}} + \frac{H}{R}\xi \\ L_{\text{Helmert}} &= L_{\text{Pizzetti}} + \frac{H}{R}\eta\sec B \end{aligned} \quad (6.14)$$

这一关系可由图 6.1 看出，由于 $QQ_0 = H\varepsilon$，$R = 6\,371$ km 是地球平均半径。即使 $\varepsilon = 1'$，且 $H = 1\,000$ m，QQ_0 之间的距离也仅有约 30 cm，大地坐标的差小于 0.01″，低于天文观测的精度，因此，大多数情况下可忽略这两种投影方式之间的差别。

由于大地水准面上点 P_0 跟椭球面上点 Q_0 之间有确切的对应关系，毕兹特投影更适于大地水准面，赫尔默特投影有实际优势，尤其是可以直接将大地坐标 B, L, h 转化为空间直角坐标 X, Y, Z，在一些其他方面也更简便。由于这个原因，我们主要采用赫尔默特投影，但实际上，结果对两种投影方式都成立。

然而，由于铅垂线的弯曲，必须仔细区分天文坐标是相对于地面点 P 还是相对于大地水准面上点 P_0。图 6.1 中的角 δ 即便只有 1 弧秒，地理纬度 1″ 的变化意味着点 P 产生的位移为 $R\delta = 30$ m。如果要将天文坐标 φ 和 λ 跟重力垂线偏差 ξ 和 η 合并到一起，一定要注意区分上面这一点，因为天文坐标是在地面点 P 测量得到的，而重力垂线偏差是用 Vening-Meinesz 公式针对大地水准面上点 P_0 计算得到的。

6.1.4 椭球参数的数值

GRS1980 包含以下四个定义参数：

$$a = 6\,378\,137 \text{ m}$$
$$GM = 398\,600.5 \times 10^9 \text{ m}^3\text{s}^{-2}$$
$$J_2 = 1\,082.63 \times 10^{-6}$$
$$\omega = 7.292\,115 \times 10^{-5} \text{ rad s}^{-1}$$

其中,a 为长半轴,GM 为地心引力常数(牛顿万有引力常数 G 乘以包含大气在内的地球质量 M),J_2 为二阶带谐因子,ω 为地球自转角速度。

由这些定义常数,可以确切地导出其他参数,例如,扁率为
$$f = 1/298.257\,222\,101$$
赤道重力为
$$\gamma_e = 9.780\,326\,771\,5 \text{ ms}^{-2}$$
DoD 的 WGS-84 包含以下参数:
$$a = 6\,378\,137 \text{ m}$$
$$f = 1/298.257\,223\,563$$
$$GM = 3.986\,004\,418 \times 10^{14} \text{ m}^3\text{s}^{-2}$$
$$\omega = 7.292\,115 \times 10^{-5} \text{ rads}^{-1}$$

IAG 计划在每四年一届的全体大会上,修改包括地球椭球参数在内的基本大地测量常数,即采用当时认为具有代表性的一列类似上面的值。另一方面,大地测量参考系并不需要采用当前的最新值。事实上,如果所有工作做得好,最终结果(三角点的空间位置,大地水准面,等等)与椭球参考系的具体选择无关(在合理的限度之内),椭球参考系的选择仅仅起中间过渡作用,对最终结果没有影响。获得与大地水准面尽可能吻合的参考椭球,这一愿望主要是出于对完美的追求。实践上更为重要的是长期稳定性,由于大量的数据是以选定的参考系为基础,因而这一参考系应当尽可能少地改变。

6.1.5 三轴椭球

三轴互相垂直且长度不等的椭球,需要三个定义参数。将三轴椭球应用于大地测量计算会引起不必要地复杂化,因而,在大地测量中,应避免采用它。

* * * * * * * * * * *

与大地水准面最为接近,且从数学上看切实可行的面是三轴椭球。很多学者估算过与大地水准面最为接近的三轴椭球参数,这样一个三轴椭球有三个互相垂直的轴,它们在地球中的定位如下:短轴与地球的主(极)惯量轴重合,长轴和中间轴都在赤道面内。这样,三轴椭球就由长轴($2a$)、短轴($2b$)、中间轴($2c$)以及长轴在赤道面内的方向来定义。

通常采用如下四个定义参数:

长半轴 a;

极扁率 f: $f = (a-b)/a$;

赤道扁率 f_e: $f_e = (a-c)/a$;

长轴的地理经度 λ_a。

由于二轴旋转椭球与大地水准面的偏差(<100 m)一般也达到这一量级,因此三轴椭球并没有使它跟大地水准面与重力场达到最佳吻合的预期;相比之下,这一复杂形状给大地测量

计算带来很多妨碍;最后,三轴椭球也不适合作为物理中的标准形状。因此,三轴椭球不适合作为参考形体,特殊应用除外。

6.2 旋转椭球的曲率及其应用

6.2.1 旋转椭球的曲率

引入空间直角坐标系(见图 6.2)。

坐标系原点位于椭球中心 O 点,Z 轴与椭球短轴重合,则椭球面的方程可以表示为

$$\frac{X^2 + Y^2}{a^2} + \frac{Z^2}{b^2} - 1 = 0 \tag{6.15}$$

引入平行圈半径

$$P = \sqrt{X^2 + Y^2} \tag{6.16}$$

作为一个新变量,将它代入式(6.15),并取微分,得到 P 点椭球面切线的斜率

$$\frac{\mathrm{d}Z}{\mathrm{d}P} = -\left(\frac{b}{a}\right)^2 \frac{P}{Z} = -\cot B \tag{6.17}$$

由式(6.15)和式(6.17),则得子午椭圆的参数方程如下

$$P = \frac{a^2 \cos B}{\sqrt{a^2 \cos^2 B + b^2 \sin^2 B}}$$

$$Z = \frac{b^2 \sin B}{\sqrt{a^2 \cos^2 B + b^2 \sin^2 B}} \tag{6.18}$$

子午圈和平行圈是旋转椭球的曲率线,因此主曲率半径位于子午平面(子午曲率半径为 M)和卯酉面内,卯酉面与子午面正交(卯酉曲率半径为 N)。

子午圈 $Z = Z(P)$ 的曲率为

$$\frac{1}{M} = -\frac{\mathrm{d}^2 Z / \mathrm{d}^2 P}{[\sqrt{1 + (\mathrm{d}Z/\mathrm{d}P)^2}]^3} \tag{6.19}$$

将式(6.17)和根据式(6.18)求得的二阶导数代入式(6.19),得到子午圈曲率半径

$$M = \frac{a(1 - e^2)}{(\sqrt{1 - e^2 \sin^2 B})^3} \tag{6.20}$$

平行圈平面(旋转椭球的斜截面)与同一切线方向的垂直面相交于 P 且交角为 B。Meusnier 定理给出了卯酉圈曲率半径

$$N = P/\cos B \tag{6.21}$$

应用式(6.18),经变形可得

$$N = a/\sqrt{1 - e^2 \sin^2 B} \tag{6.22}$$

比较式(6.20)和式(6.22),可发现 $N \geqslant M$。在两极($B = \pm 90°$),极曲率半径则为

$$c = M_{90} = N_{90} = a^2/b \tag{6.23}$$

在赤道($B = 0°$)上,有

$$M_0 = b^2/a, \quad N_0 = a \tag{6.24}$$

方位角为 A 的任意方向法截线曲率半径可根据欧拉公式计算

$$\frac{1}{R_A} = \frac{\cos^2 A}{M} + \frac{\sin^2 A}{N} \tag{6.25}$$

式中,R_A 为曲率半径,大地方位角 A 定义为:P_1 点的椭球子午面与由 P_1 点法线和 P_2 点确定的垂直面之间的夹角,A 在水平面内从北方向按顺时针方向量算。

6.2.2 椭球上某定点的平均曲率半径

有了给定纬度 B、方位角为 A 的法截线曲率半径,即

$$R_A = MN/(N\cos^2 A + M\sin^2 A) \tag{6.26}$$

为了求得纬度为 B 的一点上 R_A 的平均值,应用求函数平均值的定理,这一定理容易用图6.3来说明。曲线 $y = f(x)$ 的切线斜率表示为 $f'(x)$,这里上标($'$)表示求微分,则在点 $Q(\xi, f(\xi))$,斜率为 $f'(\xi)$。弦 PS 的斜率为 $(f(d) - f(c))/(d - c)$,如图所示,存在一点 Q,Q 点的切线斜率等于弦 PS 的斜率,且有 $c < \xi < d$,即,可以找到一个值使得

$$f'(\xi) = (f(d) - f(c))/(d - c) \tag{6.27}$$

在这一点上,定义 $f(\xi)$ 为函数 $f(x)$, $c \leq x \leq d$ 的平均值。根据定积分的定义,有

$$\int_c^d f'(x)\,dx = f(d) - f(c) \tag{6.28}$$

用 $f(d) - f(c)$ 的这一值代入式(6.27),有

$$f'(\xi) = \frac{1}{d-c}\int_c^d f'(x)\,dx \tag{6.29}$$

通过求函数在积分限从 c 到 d 上定积分的数值,可按照上式求函数 $f(x)$, $c \leq x \leq d$ 的平均值(可以忽略式(6.29)中表示求微分的上标)。

用 R 代替 $f(\xi)$,A 的积分限为 $c = 0, d = 2\pi$,则由式(6.26)和式(6.29)有

$$R = \frac{1}{2\pi}\int_0^{2\pi} f(A)\,dA = \frac{1}{2\pi}\int_0^{2\pi} \frac{MN}{N\cos^2 A + M\sin^2 A}dA = \frac{2}{\pi}\int_0^{\frac{\pi}{2}} \frac{M\sec^2 A}{1 + (M/N)\tan^2 A}dA$$
$$= \frac{2}{\pi}\sqrt{MN}\left[\arctan\left(\sqrt{\frac{M}{N}}\tan A\right)\right]_0^{\frac{\pi}{2}} = \frac{2}{\pi}\sqrt{MN}\left(\frac{\pi}{2} - 0\right) = \sqrt{MN} \tag{6.30}$$

因此,椭球面上某给定点的平均曲率半径,为该点上主曲率半径的几何平均值。

6.2.3 子午线弧长

在地图投影公式中,经常出现一个长度量,就是从 E(从赤道开始)沿子午圈到 A 点的长度,就简单数学函数来说,这个量不能用有限项公式表示。根据子午线弧素 $dX = MdB$,弧长 X 可以从理论上用积分方法求出。在实践中,这一积分必须展开成无穷级数,且可以利用如下这一点,即任何与地球形状相似的椭球体,其第一偏心率的平方 e^2 仅仅约等于1/150。为了推导 X 的公式,把 M 的表达式展开为

$$a(1 - e^2)\left(1 + \frac{3}{2}e^2\sin^2 B + \frac{15}{8}e^4\sin^4 B + \frac{35}{16}e^6\sin^6 B + \cdots\right) \tag{6.31}$$

并把 $\sin B$ 的幂用三角恒等式代换

$$\sin^2 B = \frac{1}{2} - \frac{1}{2}\cos 2B$$

$$\sin^4 B = \frac{3}{8} - \frac{1}{2}\cos 2B + \frac{1}{8}\cos 4B \qquad (6.32)$$

$$\sin^6 B = \frac{5}{16} - \frac{15}{32}\cos 2B + \frac{3}{16}\cos 4B - \frac{1}{32}\cos 6B$$

$$\vdots$$

则

$$M = a\left(1 - \frac{1}{4}e^2 - \frac{3}{64}e^4 - \frac{5}{256}e^6\cdots\right) - a\left(\frac{3}{4}e^2 + \frac{3}{16}e^4 + \frac{45}{512}e^6\cdots\right)\cos 2B + \\ a\left(\frac{15}{64}e^4 + \frac{45}{256}e^6\cdots\right)\cos 4B - a\left(\frac{35}{512}e^6\cdots\right)\cos 6B\cdots \qquad (6.33)$$

积分得到

$$X = a\left(1 - \frac{1}{4}e^2 - \frac{3}{64}e^4 - \frac{5}{256}e^6\cdots\right)B - a\left(\frac{3}{8}e^2 + \frac{3}{32}e^4 + \frac{45}{1\,024}e^6\cdots\right)\sin 2B + \\ a\left(\frac{15}{256}e^4 + \frac{45}{1\,024}e^6\cdots\right)\sin 4B - a\left(\frac{35}{3\,072}e^6\cdots\right)\sin 6B\cdots \qquad (6.34)$$

其中第一项当中的 B 必须以弧度为单位。若令 $B = \pi/2$，则所有的三角函数项为零，仅余下 B 项本身。

这就给出了从 E 到 P 四分之一子午椭圆的弧长，有意思的是，回想起单位米的最初定义，就是定义为这一长度的千万分之一。按照近来估算的地球椭球参数，计算四分之一子午椭圆的长度为 10 002 001 m。

6.2.4 平行圈弧长

介于地理经度 L_1 和 L_2 之间的平行圈弧长按照 $dS = N\cos B dL$ 给出，即

$$S = \int_{L_1}^{L_2} N\cos B dL = N\cos B(L_2 - L_1) \qquad (6.35)$$

6.3 大地线

6.3.1 法截线和相对法截线

包含一点法线和另外一点的平面与椭球相交所形成的椭球面上两点之间的连线。例如，在 A 和 B 两点之间，A 点的法截线包含 A 点的法线并且通过 B 点，B 点的法截线包含 B 点的法线并且通过 A 点。英文中亦作 plane curve, curve of normal section，以及 normal section line。

在椭球面上，有从 A 点和 B 点开始，分别到 B 点和 A 点结束的大地法截线，两个法截线并不重合。

因此，在这两点的任一点上，有两条相交的大地法截线，实际上这会造成一定的困惑，方位角应该量至哪一条法截线呢？为了避免造成混乱，引入大地线。

6.3.2 大地线

数学表面上两点之间的最短线。大地线是双重弯曲的曲线,若数学曲面是椭球面,则大地线位于相对法截线之间。英文中亦作 geodesic line。当所研究的曲面为参考椭球面时,也采用 geodesic line 这个术语。

****

为了在旋转椭球面上进行计算,椭球面上的点之间必须用椭球面曲线相互连接。这里主要讨论法截线(弧)和大地线。

法截线由垂直面和椭球的截线来定义,因此,用经纬仪观测得到的、并归算到椭球面上的方向,所夹的角即为法截线之间的夹角;斜距也可以归算为法截线的长度。由于椭球上两点的面法线一般是不共面的,故从 P_1 点到 P_2 点跟从 P_2 点到 P_1 点的相对法截线并不重合(见图 6.4)。为了得到符合定义的计算结果,必须要考虑方位角之差 $A_1' - A_1''$,若 $S = 50 \text{ km}$,则方位角之差最大仅为 $0.02''$。

通常,引入大地线是因为它在微分几何中具有良好的性质,且具有唯一性。大地线是椭球面上两点之间的最短连线,一般位于两条相对法截线之间(见图 6.4)。

由图 6.5 可以直接得到

$$\frac{dB}{dS} = \frac{\cos A}{M}, \quad \frac{dL}{dS} = \frac{\sin A}{N\cos B} \tag{6.36}$$

由这两个关系式,再根据克莱劳方程,有

$$N\cos B \sin A = \text{const.} \tag{6.37}$$

若式(6.37)对 S 求导,顾及式(6.36),得

$$\frac{dA}{dS} = \frac{\sin A \tan B}{N} \tag{6.38}$$

为了进行级数展开,式(6.36)和式(6.38)就构成了重要的大地线一阶微分方程。

针对法截线的方位角 A_1' 和弧长 S',应按下式归算为大地线的方位角 A_1 和弧长 S:

$$A_1' - A_1 = \frac{e^2}{12a^2}\cos^2 B_1 S^2 \sin 2A_1 + \cdots \tag{6.39}$$

$$S' - S = \frac{e^4}{360a^4}\cos^4 B_1 S^5 \sin^2 2B_1 + \cdots \tag{6.40}$$

若弧长 $S = 50(200) \text{ km}$,则方位角的改正量最大为 $0.007''(0.112'')$,弧长的改正量不超过 $2 \times 10^{-11}(2 \times 10^{-8}) \text{ m}$,因此后者总是可以忽略不计,方位角的改正量达到了长距离边一等角度测量的精度。高斯(C. F. Gauss)全面地研究了大地线,大地线在大地测量中的重要作用在教科书和大地测量手册中都有论述。

6.4 到参考椭球面的归算

6.4.1 水平方向归算到椭球面

在地球表面测得的方向和距离需要归算为参考椭球面上的相应值,对于方向观测需要施

加的改正包括标高差改正、截面差改正和垂线偏差改正。标高差改正为

$$\delta_2 = \frac{e^2}{2b}h_2\cos^2 B_1 \sin 2A_1 \tag{6.41}$$

截面差改正为

$$\delta_3 = -\frac{e^2}{12a^2}\cos^2 B_1 \sin 2A_1 \tag{6.42}$$

垂线偏差改正为

$$\delta_1 = -(\xi\sin A_1 - \eta\cos A_1)\cot Z_1 \tag{6.43}$$

式中,h_2 为目标点的大地高;B_1 为测站的大地纬度;A_1 为从测站到目标点方向的大地方位角;ξ 和 η 分别为天文大地垂线偏差在测站子午面和卯酉面内的分量;Z_1 为天顶距。

6.4.2 斜距归算

电磁波测距得到两点 A 与 B 之间的空间斜距(见图 6.6)。这种距离既可以直接用于大地坐标系 B,L,h 中的计算,正如在"三维大地测量学"中一样,也可以将它们归算到椭球面以获取弦长 l_0 或大地线长 s_0。

再次把椭球弧 A_0B_0 近似为半径为 R 的圆弧,R 是沿弧 A_0B_0 的平均椭球曲率半径,在三角形 OAB 中应用余弦定理,得

$$l^2 = (R+h_1)^2 + (R+h_2)^2 - 2(R+h_1)(R+h_2)\cos\psi \tag{6.44}$$

顾及

$$\cos\psi = 1 - 2\sin^2\frac{\psi}{2} \tag{6.45}$$

则有

$$l^2 = (h_2 - h_1)^2 + 4R^2\left(1+\frac{h_1}{R}\right)\left(1+\frac{h_2}{R}\right)\sin^2\frac{\psi}{2} \tag{6.46}$$

又及

$$l_0 = 2R\sin\frac{\psi}{2} \tag{6.47}$$

记 $\Delta h = h_2 - h_1$,得

$$l^2 = \Delta h^2 + \left(1+\frac{h_1}{R}\right)\left(1+\frac{h_2}{R}\right)l_0^2 \tag{6.48}$$

因而,弦长 l_0 和弧长 s_0 可表示为

$$l_0 = \sqrt{\frac{l^2 - \Delta h^2}{(1+h_1/R)(1+h_2/R)}}$$

$$s_0 = R\psi = 2R\arcsin\frac{l_0}{2R} \tag{6.49}$$

这些公式的椭球改进公式可参考 Rinner(1956)(Rinner, K. 1956. Ueber die Reduktion grosser elektronisch gemessener Entfernungen. Zeitschrift Fuer Vermessungswesen, V.81)。

6.4.3 天文观测归算到参考椭球

现在,依据赫尔默特投影来建立自然坐标 φ,λ,H 和参照于椭球的大地坐标 B,L,h 之间的

关系。暂时先不考虑高程 h 和 H，也可以把问题表述为，将天文坐标 φ 和 λ 归算到椭球上。如果还将天文方位角观测包含进来的话，为了获取大地坐标 B 与 L 以及大地方位角 A，必须把天文坐标 φ 与 λ 以及天文方位角 α 归算到椭球上。假设有一个单位球（球半径为 1），球心在测站 P，实际铅垂线与单位球交于天文天顶 Z_a，而椭球法线与单位球交于大地天顶 Z_g。图 6.7 表示出从高处看到的单位球。瞄准目标的照准线，即为该方向观测方位角 α，与单位球交于点 T，且相对于天顶 Z_a 和 Z_g 的天顶距分别为 Z' 和 Z。点 P_N 相应于北极方向，该点相对于 Z_a 和 Z_g 的天顶距分别为 $90°-\varphi$ 和 $90°-B$；Z_a 和 Z_g 相对于 P_N 的夹角为天文经度和大地经度之差 $\lambda - L$。Z_a 点的角度为天文方位角 α，相应于 Z_g 点的大地方位角 A。点 F 位于天文子午圈上，即连接 P_N 和 Z_a 的大圆，使得角 Z_aFZ_g 等于 $90°$；$\xi = Z_aF$ 和 $\eta = Z_gF$ 为垂线偏差的两个分量。首先考察以 Z_g，F 和 P_N 为顶点的球面直角三角形，根据纳白尔规则（Napier's rules），有

$$\sin B = \cos(90° - \varphi + \xi)\cos\eta \qquad (6.50)$$
$$\sin\eta = \cos(90° - (\lambda - L))\cos B$$

由于角 η 和 $\lambda - L$ 是小角度，可以利用下面的近似

$$\cos\eta \doteq 1, \quad \sin\eta \doteq \eta, \quad \cos[90° - (\lambda - L)] = \sin(\lambda - L) \doteq (\lambda - L) \qquad (6.51)$$

从而有

$$\xi = \varphi - B \qquad (6.52)$$
$$\eta = (\lambda - L)\cos B$$

这是用地理坐标（天文坐标和大地坐标）表示垂线偏差分量 ξ 和 η 的基本公式，从而将天文坐标和大地坐标联系起来。

方位角之差

$$\Delta A = \alpha - A \qquad (6.53)$$

包含两部分，$\Delta_1 A$ 和 $\Delta_2 A$（见图 6.7）

$$\Delta A = \Delta_1 A + \Delta_2 A \qquad (6.54)$$

$\Delta_1 A$ 可从球面三角形 $N_g N_a P_N$ 求取，方法很明显与前面用到的球面三角形 $Z_g F P_N$ 类似，$N_g P_N = B$ 对应于 $Z_g P_N = 90° - B$，$\Delta_1 A$ 对应于 η。因此，与 $\eta = (\lambda - L)\cos B$ 对应的方程为

$$\Delta_1 A = (\lambda - L)\sin B \qquad (6.55)$$

联合 $\eta = (\lambda - L)\cos B$，式(6.55)变为

$$\Delta_1 A = \eta\tan B \qquad (6.56)$$

在连接 Z_g 和 T 的大圆上引入一点 G，使得角 $Z_a G Z_g$ 等于 $90°$，令 $Z_a G = \delta$，可以看出图形 $Z_a G T T_g T_a$ 和图形 $Z_g F P_N N_a N_g$ 的几何构形相同，因而 $\Delta_2 A$，δ，Z' 与 $\Delta_1 A$，η，$90° - B$ 对应，故，对应于 $\Delta_1 A = \eta\tan B$ 的方程为

$$\Delta_2 A = \delta\cot Z' = \delta\cot Z \qquad (6.57)$$

由于图形 $Z_a F Z_g G$ 很小，故可视为平面图形（参见图 6.7 的放大部分），根据平面坐标转换的常用公式，有

$$\delta = \xi\sin A - \eta\cos A \qquad (6.58)$$

故

$$\Delta_2 A = (\xi\sin A - \eta\cos A)\cot Z \qquad (6.59)$$

又 $\Delta_1 A = \eta\tan B$，则最后可得

$$\Delta A = \eta\tan B + (\xi\sin A - \eta\cos A)\cot Z \qquad (6.60)$$

第一项 $\Delta_1 A$,对每个照准目标都相同,与方位角和天顶距无关;第二项 $\Delta_2 A$,与方位角和天顶距有关。第一项 $\Delta_1 A$ 的产生,是由于天文方位角是从天文北方向 N_a 而不是从大地北方向 N_g 起算的,而 N_g 是大地方位角 A 的起算方向,因而,第一项代表了零点的一个变换,这对所有方向都相同。第二项 $\Delta_2 A$ 是由于目标 T 从 Z_a 和 Z_g 投影到水平面上的不同点 T_a 和 T_g 所引起的;其影响跟经纬仪水准器不准确所产生的影响相同。

在一等三角测量中,视线通常是接近水平的,以至于 $Z = 90°$,$\cot Z = 0$,因此,$\Delta_2 A$ 项改正一般可以忽略不计,从而有

$$\Delta A = \eta \tan B = (\lambda - L)\sin B \tag{6.61}$$

这是拉普拉斯方程常见的简化形式。值得注意的是,方位角之差 $\Delta A = \alpha - A$ 和经度之差 $\lambda - L$ 以如此简单的形式联系起来。

供以后参考之用,注意到总垂线偏差(实际铅垂线和椭球法线之间的夹角 θ)可以表示为

$$\theta = \sqrt{\xi^2 + \eta^2} \tag{6.62}$$

以及,垂线偏差在方位角 A 方向的分量 ε 为

$$\varepsilon = \xi \cos A + \eta \sin A \tag{6.63}$$

这两个方程都可以直接从图 6.7 中的放大部分得到;δ 与 ε 同 ξ 与 η 之间的关系可以用平面坐标转换得到。

最后,相对于大地水准面的正高 H 与相对于椭球的大地高 h 之间的关系可以直接写出,因为,根据图 6.1(参见"投影方法"),再做充分近似,可以看出

$$h = H + N \tag{6.64}$$

因而,从自然坐标到大地坐标的转换公式为

$$\begin{aligned} B &= \varphi - \xi \\ L &= \lambda - \eta \sec B \\ h &= H + N \end{aligned} \tag{6.65}$$

相应的方位角公式为

$$A = \alpha - \eta \tan B \tag{6.66}$$

在应用这些公式时,需要相对于选定参考椭球的大地水准面差距 N 和垂线偏差分量 ξ 与 η。

有两点应当注意:

(1)参考椭球的短轴与地球自转轴平行(否则,在图 6.7 中,将会有两个不同的极 P_N),但椭球不必进行绝对定位,即不必使椭球中心与地球质心重合。

(2)垂线偏差分量 ξ 和 η 直接参考于进行天文观测的地面点,而不是大地水准面。

若是依据 Vening-Meinesz 公式,用重力测量法计算大地水准面上的垂线偏差分量 ξ 和 η,而 B,L,h 和 A 是参考于绝对定位的椭球,则应当注意铅垂线的弯曲。

还应当提到,大地方位角 A 是针对实际目标 T 的,而目标一般并不在椭球面上。对于常规椭球计算方法,人们希望方位角是针对椭球面上目标 T_0 的,T_0 是通过 T 的法线与椭球的交点。此外,A 针对的是椭球上的法截线,而不是大地线,但计算中用到的是大地线。这两种情况下,很小的方位角改正是必要的,由于这些改正是纯椭球大地测量问题,读者可参考任何一本椭球大地测量学的教科书。

6.4.4 拉普拉斯方程(方位角)及其作用

将天文与大地方位角同天文与大地经纬度联系到一起的表达式:

$$\alpha - A = (\lambda - L)\sin\varphi \quad (6.67)$$

式中，α 为天文方位角，A 为大地方位角，λ 为天文经度，L 为大地经度，φ 为纬度（方位角从北方向顺时针量算，经度向东为正）。

该方程对加强三角网的强度非常重要，因为它给出了一种控制三角网角度变形的方式。可以构成拉普拉斯方程的三角点（即观测了天文经度和方位角），又称拉普拉斯站，或拉普拉斯点。

假定在地球表面的一点 P 上，测得了到另外一点 Q 的天文方位角 α 和天顶距 Z'，α 和 Z' 都是以 P 点实际铅垂线为参考的，因为是把经纬仪的垂直轴整置成与铅垂线重合。假如，可以把经纬仪的垂直轴整置成与 P 点相对于参考椭球的法线重合，则可以测量出"大地"方位角 A 和"大地"天顶距 Z。天文方位角 α 和天文天顶距 Z' 与相应的大地方位角和大地天顶距之间的关系为

$$\alpha - A = \eta\tan B + (\xi\sin A - \eta\cos A)\cot Z \quad (6.68)$$

$$Z' - Z = -(\xi\cos A + \eta\sin A) \quad (6.69)$$

特别的，在一等三角测量中，所有的照准线通常是将近水平的，即 $Z = 90°$，则式（6.68）可简化为

$$\alpha - A = \eta\tan B \quad (6.70)$$

应用方程 $\eta = (\lambda - L)\cos B$，则有

$$\alpha - A = (\lambda - L)\sin B \quad (6.71)$$

式（6.71）称之为拉普拉斯方程或拉普拉斯条件。若在一个测站上，能够测量出天文经度 λ 和天文方位角 α，且能够独立地获取（通过三角测量计算）所对应的大地经度 L 和大地方位角 A，则式（6.71）构成了这四个量 α, A, λ, L 必须要满足的条件。显然，若 Z 不等于 $90°$，根据式（6.68）和 $\xi = \varphi - B, \eta = (\lambda - L)\cos B$，则拉普拉斯方程的确切形式为

$$\alpha - A = (\lambda - L)\sin B + [(\varphi - B)\sin A - (\lambda - L)\cos B\cos A]\cot Z \quad (6.72)$$

所有这些方程都是基于椭球坐标轴 xyz 与地球坐标轴 XYZ 的平行性导出的，因而，在三角测量计算中应用这些公式，将会起到确保这两个直角框架的平行性的作用。

为了理解这种情形，我们从几何的角度，简化地考虑如何进行三角测量计算。为了使几何结构容易观察，假定测量是没有误差的。

为了简单，首先假设三角测量的所有测站都位于参考椭球的表面（下一步，将会消除这种过度简化）；但铅垂线并不与椭球法线重合，即允许存在垂线偏差 ξ, η；给出必要的距离（空间斜距，即椭球上点间的弦长）和水平角，以确定椭球上的几何构形；在每个点上观测出天文坐标 φ 和 λ；在一个测站上，即大地原点 P_0 上，除 φ_0 和 λ_0 外还测出一个天文方位角 α_0；还给出 P_0 点的地心坐标 X_0, Y_0, Z_0。

各三角点的大地坐标 B 和 L（根据假设，大地高 h 为零）可按照下面的方法求解。根据从 (B, L, h) 到 (X, Y, Z) 的转换公式，对原点 P_0，有

$$\underline{X}_0 = \underline{X}(B_0, L_0, h_0) \quad (6.73)$$

从这几个方程解出 B_0, L_0, h_0。由于原点 P_0 位于椭球面上，由无误差数据必有 $h_0 = 0$。

有了测得的天文坐标 φ_0 和 λ_0，则可以计算出 P_0 点的垂线偏差

$$\xi_0 = \varphi_0 - B_0, \eta_0 = (\lambda_0 - L_0)\cos B_0 \quad (6.74)$$

则由式（6.68），实际上可简化为式（6.70），得

$$A_{01} = \alpha_{01} - \eta_0 \tan B_0 \tag{6.75}$$

由测量得到的天文方位角 α_{01} 确定了初始大地方位角 A_{01}。

在 P_0 点测得的水平角也可以归算到椭球面上

$$\gamma_{12}^{\text{ell.}} = \gamma_{12}^{\text{meas.}} + (\xi_0 \sin A_{01} - \eta_0 \cos A_{01}) \cot Z_{01} - (\xi_0 \sin A_{02} - \eta_0 \cos A_{02}) \cot Z_{02} \tag{6.76}$$

这是通过对两个类似式(6.68)的方程取差得到的。主项 $\eta \tan B$ 消掉了,结果是,对于接近水平的照准方向,所夹水平角的归算量非常小,通常可以忽略不计。

现在,可以计算从 P_0 点起始的所有方向 $P_0 P_1$,$P_0 P_2$,… 的方位角(见图6.8)

$$\begin{aligned} & A_{01} \text{ by } (6,75) \\ & A_{02} = A_{01} + \gamma_{12} \\ & A_{03} = A_{02} + \gamma_{23} \end{aligned} \tag{6.77}$$

式中,γ_{ij} 表示椭球面上的角度值 $\gamma_{ij}^{\text{ell.}}$。

通过众所周知的椭球计算方法,应用弦长可以计算出 P_0 点周围各个三角点(图6.8中的 P_1,P_2,P_3,P_4)的大地坐标,以及反方位角 A_{10},A_{20},A_{30},A_{40}(无需任何额外的方位角测量!)。

下一个点,比如说 P_1,可以用完全相同的方法处理。前面的椭球计算已经给出了 B_1,L_1,A_{10};由天文测量已知 φ_1,λ_1,故可给出 ξ_1,η_1;然后,可将在 P_1 点测得的水平角归算到椭球上(如果必要的话),再继续计算 P_1 点周围的三角点,依此类推。

用这种方法,获取了所有三角点的大地坐标 B,L,以及所有边的大地方位角 A。

很重要的一点在于,完全确定三角网的定向,仅需一个用天文测量得到的方位角(这里是 α_{01}),而椭球 Z 轴和地球 Z 轴的平行性是通过选择大地基准来保证的。

假如又测得了另一个天文方位角,比方说 α_{56},情况将会怎样?根据前面的讨论,由于已知 φ_5,λ_5,B_5,L_5,所以可以求 ξ_5 和 η_5,从而可以根据式(6.68) 计算 A_{56}。但是,根据前面的椭球计算已经求出了 A_{56},因此,A_{56} 的这两个值,一个是根据椭球计算得到,一个是将测量出的 α_{56} 归算到椭球上得到,必须是吻合的,换言之,必须要满足拉普拉斯条件式(6.71)

$$(\alpha_{56} - A_{56}) = (\lambda_5 - L_5) \sin B_5 \tag{6.78}$$

A_{56} 是根据椭球计算得到的数值(更确切地说当然是式(6.72))。

如前面所做的假设,若观测是无误差的,则这一拉普拉斯方程自然会得到满足,因为几何条件是唯一确定的。若观测受到测量误差的影响,则拉普拉斯条件将不再满足,但可以通过平差计算来强制实现。附加一个天文方位角观测量,则给出了用于平差的一个拉普拉斯条件。这样,理想的几何条件(精确的定向以及坐标轴的平行性)就能在最可能好的程度上重建。

在这种意义上,论断"拉普拉斯条件确保网的定向以及坐标轴的平行性"应理解为:它是一个条件,不是为了固定几何条件,而是为了调整测量误差。

迄今为止,所做的讨论都是假定三角点是位于椭球面上的。现在,去掉这一假设,即认为三角点在地球的表面上。因此,需要关于点位高程的额外信息,有两种主要处理方法:天文大地测量法,应用天顶距的三维三角测量法。

理论上,第二种方法更有吸引力,但在采用天顶距时遇到了实际困难,因此,在实际中首选第一种方法。

在天文大地测量方法中,用天文水准法确定大地水准面高,再结合正高,用来把这种情况转化为第一种简单情况,就是通过迭代的过程,归算为点在椭球面上的情况。相对于拉普拉斯方程的几何构形实质上保持不变。

在三维三角测量中,可以把观测量(方位角、水平角、天顶距)归算为以椭球法线为准,并采用大地坐标。这样,这种情形也同样转化为第一种情况,可通过赫尔默特投影投影到椭球上的三角点,扮演了前面位于椭球上点的角色。同样,相对于拉普拉斯方程的几何构形保持不变。需要一个天文方位角确定几何定向,其他方位角帮助调整测量误差,由此产生的几何条件尽可能地接近理想状况:正确的定向和平行性。

如果在下面的情况下,由于几何的需要而测量了更多的天顶距,则多余的观测量给出了形如式(6.69)的条件。在这个意义上,就可以理解式(6.68)和式(6.69)两个方程构成的系统近来被称为"扩展拉普拉斯方程"的原因了。但是,天顶距的这种条件仅仅有助于改进高程,对于定向和坐标轴没有实际影响。因此,本书作者更喜欢保留方位角条件的"拉普拉斯方程"这个名称。

有人询问,为了确定几何定向需要多少必要的拉普拉斯条件。在作者看来,这个问题不应当这样提,因为它是模棱两可的:答案可以是从零(只考虑多余的条件)到任意数目(如果是把任何水平角和垂直角观测量归算看作是"扩展拉普拉斯方程"式(6.68)和式(6.69)的应用的话)。要是问固定几何图形所必要的观测量数目的话会显得更清楚:每一点上的 φ 和 λ,足够的水平角、距离以及(根据具体情况而定)天顶距用来确定几何构形,一个天文方位角用来定向。总数为 n 的天文方位角(在 n 个不同点上)给出 $n-1$ 个拉普拉斯条件,从几何上讲是多余的,但实际当中非常重要。这样的方位角越多、分布越好,控制网的定向以及坐标轴的平行性就越好。

6.4.5 椭球面三角形解算

为了利用距离和方位角确定地理坐标,没有观测的边长和角度必须要通过椭球面三角形来计算。对于一等三角测量的边长,椭球面上的计算可以用球面上的计算来代替,从这一角度,椭球可近似为高斯密切球,与椭球切于 $P_0(B_0)$,球半径为

$$R_0 = \sqrt{M_0 N_0} \tag{6.79}$$

切点处纬度 B_0 可以选为特定三角形各顶点地理纬度的算术平均数。在高斯密切球切点周围半径为 150 km 的范围内,由球近似引起的方向误差不超过 $0.005''$。

为了解算球面三角形,若球面三角形的边长为 a,b,c,内角为 A,B,C,则根据球面正弦定理,有

$$\frac{\sin A}{\sin B} = \frac{\sin(a/R)}{\sin(b/R)} \tag{6.80}$$

将 $\sin(a/R)$ 和 $\sin(b/R)$ 展开成直到 $O(1/R^3)$ 的级数式,可得到与平面正弦定理相应的关系式,若将球面三角形的内角减去三分之一球面角盈(球面角超),则可得到勒让德方程

$$\frac{\sin(A-E/3)}{\sin(B-E/3)} = \frac{a}{b} \tag{6.81}$$

若球面边长的级数增加一项,则结果就是 Soldner 增量法

$$\frac{\sin A}{\sin B} = \frac{a - a^3/6R^2}{b - b^3/6R^2} \tag{6.82}$$

式(6.81)中出现的球面角盈 E 是球面三角形内角和超出 180° 的部分,根据下式计算

$$E = F/R^2 \tag{6.83}$$

式中，F 为用球面边长计算出的平面三角形面积。边长 $S = 50$ km 的等边三角形，角超 $E = 5.48''$。

6.4.6 勒让德定理

在测量计算中，经常需要做的一项运算就是解算三角形，这牵涉到三角形中边和角之间的关系。

测量员关心三角形中各边的长度，而不是这些边在球心所代表的角度。在早期大地测量计算中，把三角形当作球面三角形处理，把边的长度用合适的球半径转化为以角度为单位，计算出来的以角度为单位的边长再转换回以长度为单位，这明显是一个冗长的过程。是勒让德提出了另外一种解法，把该解法作为一个练习题写在他 1813 年出版的一本关于三角学的书中。

勒让德定理表明，事实上，若球面三角形与球的大小相比很小的话，则球面三角形可认为由三条绳子构成，且可以从球面上移去并拉平，由此得到的平面三角形三内角与球面三角形三内角之差几乎完全相同，该差值为 $\frac{1}{3}E$。这个定理表明，若已知球面三角形一条边的长度，三内角各减去 $\frac{1}{3}E$，则下面的关系

$$\frac{a}{\sin\left(A - \frac{1}{3}E\right)} = \frac{b}{\sin\left(B - \frac{1}{3}E\right)} = \frac{c}{\sin\left(C - \frac{1}{3}E\right)} \tag{6.84}$$

近似成立，可用于大地测量方面的三角形解算，用来计算三角形中其他两条边的长度。

该定理的证明概述如下，基于拉直三角形思想的另一种证明方法，可参考其他书。如图 6.9(1) 所示的球面三角形，像往常一样，边用角度为单位表示，为 α, β, γ，对应长度为 $\alpha R, \beta R, \gamma R$，$R$ 为球半径。考虑如图 6.9(2) 所示的一平面三角形，各内角的角度如图示，内角之和自然是 180°，可得

$$\frac{x}{y} = \frac{\sin\left(A - \frac{1}{3}E\right)}{\sin\left(B - \frac{1}{3}E\right)} = \frac{\sin A - \frac{1}{3}E\cos A \cdots}{\sin B - \frac{1}{3}E\cos B \cdots} \tag{6.85}$$

现在，来计算球面角盈，其数值很小，球面三角形的面积可表示为

$$\frac{1}{2}\beta R \gamma R \sin A, \text{ 或 } \frac{1}{2}\gamma R \alpha R \sin B, \text{ 或 } \frac{1}{2}\alpha R \beta R \sin C$$

故，应用公式（三角形的球面角盈 $E = (\text{triangle area})/R^2$），可得

$$E = \frac{1}{2}\beta\gamma\sin A = \frac{1}{2}\gamma\alpha\sin B \quad \left(= \frac{1}{2}\alpha\beta\sin C, \text{also}\right) \tag{6.86}$$

出于同样的原因，分式 x/y 中出现的余弦可用下面的公式代换

$$\cos A = \frac{\beta^2 + \gamma^2 - \alpha^2}{2\beta\gamma} \text{ and } \cos B = \frac{\gamma^2 + \alpha^2 - \beta^2}{2\gamma\alpha} \tag{6.87}$$

这是根据平面三角学导出的。做代换，可得

$$\frac{x}{y} = \frac{\sin A \left[1 - \frac{1}{12}(\beta^2 + \gamma^2 - \alpha^2)\right]}{\sin B \left[1 - \frac{1}{12}(\gamma^2 + \alpha^2 - \beta^2)\right]} \quad (6.88)$$

又

$$\frac{\sin A}{\sin B} = \frac{\sin \alpha}{\sin \beta} = \frac{\alpha - \frac{1}{6}\alpha^3 \cdots}{\beta - \frac{1}{6}\beta^3 \cdots} = \frac{\alpha\left(1 - \frac{1}{6}\alpha^2 \cdots\right)}{\beta\left(1 - \frac{1}{6}\beta^2 \cdots\right)} \quad (6.89)$$

代换 $\frac{\sin A}{\sin B}$，且在级数中只保留二阶项，可得

$$\frac{x}{y} = \frac{\alpha}{\beta} \cdot \frac{1 - \frac{1}{12}(\beta^2 + \gamma^2 + \alpha^2) \cdots}{1 - \frac{1}{12}(\gamma^2 + \alpha^2 + \beta^2) \cdots} \quad (6.90)$$

级数消去二阶以上的项，只保留 $\frac{x}{y} \approx \frac{\alpha}{\beta}$，表明在提到的精度内，平面三角形的边长跟球面三角形的边长成比例。

若对公式做进一步的研究，可以看出，用勒让德定理引起的差异小得令人吃惊：若三角形大到能够覆盖整个英国，比值 $x/\alpha, y/\beta$ 和 z/γ 的差别不超过百万分之二！

必须要强调的是，勒让德定理的应用仅仅是作为计算的辅助，跟球面三角形相联系的勒让德三角形并不能从几何上通过简单构造而联系起来。

****

若球面三角形的边长与球半径相比很短，且存在一个平面三角形，各边的长度等于相应球面三角形的各边，则两个三角形对应角度之差近似相等，即等于三角形球面角盈的三分之一。

6.5 大地主题解算的正解与反解

6.5.1 归化纬度

球与椭球在大地赤道处相切，过椭球面上需要定义归化纬度的一点，做垂直于赤道面的直线，与球交于一点，球上这点对应的半径与赤道面之间在球心的夹角。归化纬度是在大地测量学和制图学问题中使用的一种辅助纬度。在天文工作中，使用术语归化纬度时，表示地心纬度。英文中亦作 geometric latitude，parametric latitude。

为了计算使用，在图 6.10 中引入两种特殊的纬度（归化纬度和地心纬度），图中给出了 P 点的子午面。线 OP 为 P 点的地心半径，角 φ'，即该半径和大地赤道之间的夹角，为地心纬度；角 u，线 OP'' 与赤道之间的夹角，称为归化纬度。点 P'' 是通过把 P 沿平行于自转轴的方向投影到半径为 a 的辅助圆上而得到的；点 P''' 是通过把 P 沿平行于大地赤道的方向投影到半径为 b 的辅助圆上而得到的。三种纬度之间的关系为

$$\tan\varphi' = \frac{b}{a}\tan u = \frac{b^2}{a^2}\tan B$$
$$\tan\varphi' = (1-f)\tan u = (1-f)^2\tan B \tag{6.91}$$
$$\tan\varphi' = \sqrt{1-e^2}\tan u = (1-e^2)\tan B$$

地心半径的长度 ρ 为

$$\rho = \frac{a\sqrt{1-e^2}}{\sqrt{1-e^2\cos^2\varphi'}} \tag{6.92}$$

6.5.2 大地主题解算的正解与反解

大地测量计算中的一个基本问题,就是欧洲大陆作者一般所指的大地主题解算,指的是,给定某点从一已知坐标点起始的方位和距离,计算该点的椭球纬度和经度。例如,主要三角网系统经过平差取得内部一致性之后,需要计算参考椭球上点的位置,得到用于进一步加密测量或制图的控制网。

为了把系统定位在椭球上,给定一个点的大地坐标和一条边的方位就足够了,其他点的坐标则可以用适合于椭球面几何的公式,经精确计算得到。

更具体地说,点 P_1 在参考椭球上的纬度为 B,经度为 L,在 P_1 点到另外一点 P_2 的方位角为 A,从 P_1 点到 P_2 点的距离为 S,则 P_2 点的经纬度是多少呢？解决这个问题有几套良好的公式。考虑需要什么计算精度是必要的,显然这必须跟平差后测量可能达到的精度联系起来。若 40 km 边长的方位可精确到 $(1/2)''$,则两端点之间的相对定位精度为 0.1 m,或换算为相对精度 1/400 000。用 EDM 技术精密测量长度也能达到这个精度。纬度为 $1''$ 的子午线长度约为 31 m,故纬度计算到 $0.001''$ 对应 0.03 m。

因此,为了避免大规模精密计算中误差的累积,长度可取至 0.01 m,角度取至 $0.01''$,纬度和经度取至 $0.000\ 1''$,但使用接近 $0.1''$ 的角度和方位角的最终值、或 $0.001''$ 的最终坐标值,就没有作用了。

* * * * * * * * * * * *

在椭球计算中,再三出现下面的问题:

(1) 给定 P_1 点的坐标 B_1, L_1,方位角 A_1 以及距离 S,计算 P_2 点的地理坐标 B_2, L_2,以及方位角 A_2;

(2) 给定 P_1, P_2 点的坐标 B_1, L_1, B_2, L_2,计算方位角 A_1, A_2 和距离 S。

这些问题分别是大地主题解算的正解和反解。由于它们在三角网设计中的重要性,在德国的文献中,将这些问题称为第一和第二大地主题解算。无论哪个问题,都牵涉到椭球面极三角形 P_1NP_2(见图 6.11)的解算。若引入大地线作为 P_1 和 P_2 之间的椭球面曲线,则上面的问题要求对大地线微分方程积分并求解需要的量。众多的解法可分为三类。

第一类解法以对大地线微分方程系统积分为基础,微分方程系统,即,$dB/dS = \cos A/M$,$dL/dS = \sin A/N\cos B$ 和 $dA/dS = \sin A\tan B/N$。这里,勒让德(1806)将纬差、经差和方位角差作为弧长的函数进行泰勒级数展开

$$B_2 - B_1 = \left(\frac{dB}{dS}\right)_1 S + \frac{1}{2}\left(\frac{d^2B}{dS^2}\right)_1 S + \cdots$$

$$L_2 - L_1 = \left(\frac{dL}{dS}\right)_1 S + \frac{1}{2}\left(\frac{d^2L}{dS^2}\right)_1 S + \cdots \quad (6.93)$$

$$A_2 - A_1 = \left(\frac{dA}{dS}\right)_1 S + \frac{1}{2}\left(\frac{d^2A}{dS^2}\right)_1 S + \cdots$$

应用一阶导数,也可以计算更高阶的导数。由于级数是相对于 S 展开的,收敛较慢,所以一般只适用于直到 150 km 的距离。Dorrer(1966)按照龙格－库塔法(Runge-kutta method)对微分方程进行数值积分。

在 Bessel(1826)和 Helmert(1880)研究的方法中,将椭球面极三角形转换到一个同心球上(半径为 a),然后,计算就在球上进行,随后再转换回椭球上。用归化纬度 u 作为球面纬度,u 跟 B 的关系为 $\tan u = (b/a)\tan B$。若将方位角 A_1 转换到球面上,则根据克莱劳方程(Clairaut's equation)($N\cos B\sin A = $ 常数,即 $\cos u\sin A = $ 常数),椭球上的方位角在转换当中保持不变。椭球面距离 S 和球面距 σ 之间,以及椭球面经差 ΔL 和球面经差 $\Delta\lambda$ 之间的关系,通过联合下面两组方程导出

$$du/d\sigma = \cos A, \quad d\lambda/d\sigma = \sin A/\cos u \quad (6.94)$$

这组方程在球面上成立,相应的椭球面方程为 $dB/dS = \cos A$ 和 $dL/dS = \sin A/N\cos B$。经过运算,可得下面的微分方程

$$dS = a\sqrt{1 - e^2\cos^2 u}\,d\sigma$$
$$dL = \sqrt{1 - e^2\cos^2 u}\,d\lambda \quad (6.95)$$

由此得到的椭圆积分,既可以通过把方根展开成级数然后逐项积分而求解,也可以通过数值方法求解。由于只是求解椭球面元素和球面元素之间较小的差值,求解过程具有很好的收敛性,因而也适合于涉及长距离大地线的计算。

最后,大地主题解算可以这样求解,先正形投影到球面,然后进行球面解算,再转换回椭球面。利用高斯密切球($R_0 = \sqrt{M_0 N_0}$)可得到简单的转换方程。由于投影引起变形,故方位角和距离要进行改正。这种方法也适合长距离。在有限的区域内,也可以利用封闭公式用于坐标的计算。

6.5.3 在可编程便携计算器上用逐点累积法求解椭球上大地主题解算正反解

一般将参考椭球上的两种基本大地测量计算叫做大地主题解算的正解和反解。大地主题解算的正解就是已知从一给定原点到某点的距离和方位角,计算该点的纬度、经度和反方位角;大地主题解算的反解是计算两已知地理位置点之间的距离和正反方位角。

为了求解参考椭球面上的大地主题解算的正反解,有超过 50 种不同的解算公式可供选择,不同方法通常以方法研究者的名字命名。若需要 1 cm 的精度,这些方法中的大多数都不能用于解算超过 150 km 的距离,有一些方法可用于解算直到 400 km 的距离,但只有少数几种方法才可用于从 400 km 到 20 000 km 的长距离大地主题解算,而 20 000 km 是地球表面两点之间的最大可能距离。

几乎所有这些方法都是从同样的椭球面大地线弧素微分方程出发的。求解的经典过程

是,用勒让德(Legendre)级数、高斯(Gauss)中纬度级数、或贝塞耳—赫尔默特(Bessel-Helmert)级数进行级数展开。根据级数展开公式、应用代换、投影到辅助面、级数项积分等因素的变化,得到了上面提到的这50多种方法的大多数。但是,可以对初始微分方程直接积分,1971年,Kivioja发表了将这种方法实际应用于数字计算机的文章。

承认很多用户不能很容易地使用计算机,并认识到手持式可编程计算器的发展和性能,本文进一步发展了 Kivioja 的早期工作,使可编程便携式计算器从根本上代替数字计算机。程序的用户只需要输入大地主题解算的已知数据:大地主题解算的正算的地理坐标、距离和方位角;大地主题解算的反算的两个点的地理坐标;再加上不同情况下参考椭球的两个半径。这种方法在 HP-67 可编程便携式计算器上编程实现了,开发的这些算法可以在任何容量和程序存储器与 HP-67 相等的计算器上运行。该程序也可在 HP-97 和 HP-41C 计算器上使用。

在大地主题解算的正算中,这种用积分法求解的方法称为逐点累积法,是把给定的距离分成 n 等份,每段长为 Δs,使得 $s = n \cdot \Delta s$。每一段 Δs 的方位角都由大地线克莱劳方程约束,以使每一段都位于大地线上,程序计算每一段的纬度和经度增量,并将增量累加到起始点的地理坐标上。同在一般的数值积分中一样,通过降低大地线每一积分段 Δs 的长度,能够增加结果的精度。

这里所说的大地主题解算的反解的解法是,近似计算距离和一个方位角,然后从一个已知点开始,利用大地主题解算的正解中用到的同样方法,计算第二个点的坐标。计算出来的坐标一般不会与给定的坐标完全相同,但计算值和已知点坐标的差值可以用来改正起初距离和方位角的近似估值。

程序的计算时间由大地线上的积分数目决定。对于大多数问题和所有的 EDM 边,一般需要5分钟或更少。在 HP-67 上,11 秒钟之内,能完成所有跟一个积分段 Δs 有关的计算。在大地主题解算的正解或反解中,要得到厘米级精度,每一段的长度 Δs,可以由用户设置,不应超过 4 km。非常长的距离上,可以让计算器整夜运行,得到最高的计算精度。

7 地图投影

7.1 地图投影概述

大多数测量活动的最终目标就是地图制图。地图是椭球曲面的平面表现,在椭球上用纬度和经度定义位置。因此,地图的绘制需要把椭球坐标转换到平面系统,这种转换即为地图投影。

但是,不必要求出每个测量点的椭球坐标,然后再将它们转换到平面系统,通常的程序是计算主测量控制网的椭球纬度和经度,可以使用部分或全部这些坐标,与天文观测一道,用来研究地球的形状和大小,其余的点只是需要用来控制测图、工程测量等。然而,由于所有的测量观测必须在地球上进行,因此,有必要保证所有点的位置都正确地表示在平面系统中。现在必须要考虑两个问题:(1)利用地图投影的几何条件和数学公式,将以纬度和经度表示的椭球坐标转换到用于平面制图的直角坐标系中;(2)测量人员可以采用这样的方法,直接从地球上的观测量获得正确的测量点平面坐标,从而绕过点在参考椭球面上的坐标,不过当然要考虑到定位点位的投影条件。

有很多种投影方法可用于将椭球面表示在平面上,且这些方法中的许多都可以在地图集中看到。由于地图集地图必须包含地球的大范围地区,或者可能是整个地球,所以,在任何平面表示中形状变形和长度变形是不可避免的,这一点是很明显的。另一方面,测量员和工程师关心的是比例尺地图,代表性的有,1/50 000 的地形图,或者 1/2 000 的地籍和工程规划图,这些通常是基于地区或国家投影系统的,限制在椭球上一个很小的范围,举例来说,比如英国这样大的区域。

如果在某一坐标系中投影覆盖的区域足够小,则容易设计一种投影方式,使得前面提到的变形察觉不到,其实是使印刷在普通纸上的地图的变形在实际当中量不出来,这些投影性质在图形上看不见,只是保留在该投影所对应的数学公式和计算方法中。

假设半径为 R 的圆球上有一角半径为 θ 的小圆,在圆自身平面内其直径为 $2R\sin\theta$,但在球面上横穿圆的距离为 $2R\theta$(用弧度)。若将圆进行投影,使得其圆周以一定的比例保持长度不变,则在地图上测得的圆直径将会很小,设投影比例为 1,直径换算成比率为 $(\sin\theta)/\theta$,即近似为 $1-(1/6)\theta^2\cdots$。现在假设需要地图的长度变形不超过 1/2 000,则 θ 必须小于 $\sqrt{6/2\,000}$,约为 1/18 弧度,约 3°。这表明,一个投影系统覆盖范围的限制,给出了对长度投影变形的限制。

****

本质上,地图是部分或全部地球表面的一种平的,或平面的表示。制图的基本问题是,不可能把一个双重弯曲的面,比如球或椭球,无变形地展开在平面上。设计不同的地图投影以保持椭球表面的某些属性不被扭曲,根据所设计地图的用途来确定保持何种属性为"真"或不被扭曲,有以下几种可能:

(1)椭球面上的各种区域与投影平面上的面积比值保持不变——等积投影;
(2)椭球面上的距离投影后无变形——等距投影;
(3)微小形状在投影时保持相似——正形投影。

保持上面这些属性中的一种,是以其他属性的变形为代价的。例如,为了在等积投影中保持面积的合适表示,形状必然要扭曲。

投影也常根据所用投影面的类型进行分类。三种常用的投影面是平面、圆锥面和圆柱面。若沿圆锥和圆柱的母线剪开,则圆锥面和圆柱面可以无变形地展成平面。这些面与椭球面相切或相割,常通过选择切线、割线或切线、割线上的点,使得感兴趣的区域接近投影中心。

历史上,曾用简单地几何投影,将椭球面投影到投影面上来生成地图,投影面上的投影点控制了相应地图上的许多性质。在现代地图中,椭球坐标和相应地图坐标之间的关系是完全解析的。

对于测量人员,最重要、最有用的地图投影是正形投影。在正形投影中,微小图形保持相似,这意味着从一点出发所有方向的长度比与方位角无关,在短距离上恒定。如此,则一点上的角度关系不变,即从正形投影中量出的角度,或绘到正形投影上的角度,跟地球表面量出的或设置的角度相同。常用的正形投影有,横墨卡托投影和兰勃特正形圆锥投影,在后面的章节中会探讨这些投影。地图坐标构成了平面直角坐标系或格网坐标系,在格网坐标系中计算比在椭球面上要简单得多,因为可以应用简单的欧氏几何,而不是利用大地坐标和椭球方位角及距离之间的复杂计算公式来完成。

基于地图投影格网坐标的应用,对于有限区域内地籍测量和工程测量的内业标图非常方便。可以在投影平面上将平行圈和子午圈绘成格子线,由于投影变形,这种格子不会是直角的。

因此,使用直角格网坐标系,而不是经纬度,对于在地图上描绘点位位置,将会方便得多。有限区域内的地籍测量、工程测量或其他水平测量,从控制网中的大地点起始并闭合到大地点,则可采用平面三角学,通过相对简单的运算在格网上实施内业计算。更进一步,为了与控制网成为一体,应用两个系统之间的数学关系,还可将每一个格网坐标转换为参考椭球上的经纬度坐标。很长时间以来,为了容易地量算坐标,在军用地图上使用格网坐标是必不可少的。

若所有的格网坐标都为正数,既方便使用,又不容易出错,为了实现这一点,通常在投影范围以南或以西指定一个假定原点,这样,在投影范围内,所有的东向和北向坐标都会是正值。

7.2 投影方程

概括地讲,地图投影可以用一对公式表示

$$N = f_1(B, L), \quad E = f_2(B, L) \tag{7.1}$$

把直角坐标,即东向坐标和北向坐标,表示为经度和纬度的函数。在某些关于测量和地图投影的文献中,坐标用 x 和 y 来表示,许多测量人员已经习惯了用 x 坐标表示在南北方向,但可能会带来混淆,甚至发生在不同测量员之间,故最好使用 N 和 E 表示,例如,在英国陆地测量部,现在就是使用这种表示方法。

在地图投影中,涉及的基本数学运算有:
(1) 把椭球坐标转换为格网坐标, $(B, L) \rightarrow (E, N)$;
(2) 把格网坐标转换为椭球坐标, $(E, N) \rightarrow (B, L)$;
(3) 将观测方位角、角度、方向和距离归算为格网平面上的相应量。

在许多参考书中已给出每种地图投影(如,横墨卡托投影和兰勃特投影)的转换和归算的公式,主要是介绍具体投影的手册。

等量纬度

为了能够尽可能直接地使用由高斯-克吕格投影导出的公式,引入椭球上的一个特殊参数 q,称为等量纬度,既方便又实际。首先考察椭球上的微分长度 dS,有

$$(dS)^2 = (MdB)^2 + (rdl)^2 = r^2\left[\frac{M^2}{r^2}(dB)^2 + (dl)^2\right] \tag{7.2}$$

定义新参数的微分为

$$dq = \frac{M}{r}dB \tag{7.3}$$

则式(7.2)可重写为

$$(dS)^2 = r^2[(dq)^2 + (dl)^2] \tag{7.4}$$

等量纬度 q 仅依据大地纬度 B 进行明确定义,且可以通过积分式(7.3)求得

$$q = \int_0^B \frac{M}{r}dB = \int_0^B \frac{M}{N\cos B}dB = \ln\tan\left(\frac{\pi}{4} + \frac{B}{2}\right)\left(\frac{1 - e\sin B}{1 + e\sin B}\right)^{e/2} \tag{7.5}$$

椭球正形投影必然要满足下面的方程(参见"横墨卡托投影中用到的公式")

$$\frac{\partial N}{\partial q}\cdot\frac{\partial N}{\partial l}+\frac{\partial E}{\partial q}\cdot\frac{\partial E}{\partial l}=0, \quad \left(\frac{\partial N}{\partial q}\right)^2+\left(\frac{\partial E}{\partial q}\right)^2=\left(\frac{\partial N}{\partial l}\right)^2+\left(\frac{\partial E}{\partial l}\right)^2 \tag{7.6}$$

也就是

$$\frac{\partial N}{\partial q}=\frac{\partial E}{\partial l}, \quad \frac{\partial E}{\partial q}=-\frac{\partial N}{\partial l} \tag{7.7}$$

7.3 正形投影

在某一地图投影中，投影表面任一微小区域内的形状保持不变，且任一点周围的所有角度无变形，换言之，椭球上两条线的交角，投影后，两条线的投影仍以同样的角度相交，将具有这种性质的投影称为正形投影，英文中亦作 orthomorphic map projection。

这是一个非常有用的属性，因为它使得这种投影的实际应用很简便。

7.3.1 兰勃特投影

兰勃特投影是有一条或两条标准纬线的正形圆锥投影。其几何解释是，圆锥或者与椭球切于一条平行圈，或者割出两条平行圈。两条标准纬线的情形类比于横墨卡托投影的椭圆柱小于椭球的情况。圆锥的顶角可以选择，以便使圆锥与椭球在要制图的区域相割(见图 7.1)。

圆锥的轴与地球自转轴重合，圆锥的顶可以在地球的北方或南方，这根据制图区域是在北半球还是在南半球而定。当把圆锥展开到平面时，坐标格网和经纬格线的关系如图 7.2 所示。平行圈投影为凹向最近极点的圆弧，而子午圈投影为向最近极点会合的直线，子午线和平行圈的交角为直角。

通常选择标准纬线，将投影区域的南北范围分成三部分，近似比例为 1/6,2/3,1/6(见图 7.2)。

长度比在东西方向为常数，但在南北方向是变化的，沿标准纬线长度比为一，在标准纬线之外长度比很大，在标准纬线之间则变小。对于典型的兰勃特投影，在南北方向上长度比因子的变化情况如图 7.3 所示。

7.3.2 横墨卡托投影

横墨卡托投影属于等角圆柱投影，可以形象化为一个包围住地球的圆柱，其定向使得轴线在赤道面内。圆柱半径通常比地球半径稍小，与地球相交于两个椭圆，这两个椭圆与中央子午线距离相等，且与中央子午线平行(见图 7.4)。

将圆柱展开到平面上，则坐标格网和经纬线格网的关系如图 7.5 所示。

子午线和平行圈的交角为直角，中央子午线为直线，其附近的子午线接近于直线(稍凹向中央子午线)，平行圈为凹向最近极点的曲线。

子午线之间的间距随远离中央子午线的距离而增加，因而长度比也随远离中央子午线的距离而增大。为了保持正形性，南北方向的长度变形应等于东西方向的长度变形。选择圆柱的半径，以使得地图图幅范围内长度变形控制得最小。当选定的圆柱小于椭球时，中央子午线上的长度比很小。长度比随远离中央子午线的距离而增加，通常将它设计成这样，在两条接近平行的截线(截线椭圆)上长度比唯一，且截线之间距离大约是截线和投影边界之间距离的三分之二。在投影边界上,长度比很大。图7.6表现出典型横墨卡托投影中各点长度比因子与横向坐标之间的关系。长度比因子是投影平面上一点的微分距离与椭球面上相应距离的比。中央子午线上长度比因子的数值,即中央长度比因子,跟投影的宽度或东西范围,以及需要的精度有关。通用横墨卡托投影(UTM)的投影带为经度6°宽,且取中央长度比因子为0.9996。在美国,很多州使用横墨卡托投影作为州平面坐标系,这些投影中,投影带的宽度不同,且中央长度比因子从0.9999到单位1不等。

通用墨卡托投影根据最大允许的长度变形来限制其东西投影带范围,且在北半球纬度限于到84°,南半球限于到80°。极区使用极平面投影。

****

横墨卡托投影正式用于英国、埃及、瑞典、波兰、葡萄牙、俄罗斯、保加利亚、芬兰、德国、南斯拉夫、挪威、英属非洲殖民地、南非、澳大利亚、美国陆军地图局以及美国多个州的平面坐标系。

现在,横墨卡托投影比兰勃特正形圆锥投影或其他投影,在大地测量计算中的应用都广泛。

7.3.3 高斯—克吕格投影

按照与墨卡托投影相同的方式建立起来的地图投影,但其投影柱面与椭球切于子午圈,而不是切于赤道。该投影主要用于南北方向的小范围地区,用于所有英国陆地测量部的地图。在这种投影中,恒向线是弯曲的。

横柱正形投影,现在通常称之为横墨卡托投影,也称作兰勃特柱正形投影,以及其他名字。

7.4 横墨卡托投影

7.4.1 横墨卡托投影中用到的公式

横投影用于经度范围较小区域的制图,在通用墨卡托投影(UTM)中,是在中央子午线两侧各3°的范围,中央子午线投影为直线。横墨卡托投影的一般投影条件是,中央子午线的长度比为一,纵向和横向长度坐标的公式使得投影是正形的。

假设 B 为某点的椭球纬度,l 为该点从中央子午线算起的经度,则 l 实际上是一个小角,不超过3°,即约1/20弧度。将横向和纵向坐标的基本公式表示成 l(弧度为单位)的幂级数。在进入集中讨论数学公式之前,对公式提以下几点:

(1)当 l 等于零时,横向坐标 E 的公式必为零,纵向坐标 N 的公式必简化为 $N = S$,其中 S 为沿子午线的直线距离,从赤道量起或者从子午线上任一指定点量起(在英国陆地测量部的地图投影中,设定的零点为北纬49°);

(2)由于投影关于中央子午线几何对称,故 l 变为 $-l$ 时,纵向坐标 N 必不变,且横向坐标 E 的公式给出的坐标数值不变,仅改变符号。因此,可以写出以下公式

$$N = S + Pl^2 + Ql^4 + Rl^6 + \cdots$$
$$E = Al + \bar{B}l^3 + Cl^5 + \cdots \tag{7.8}$$

在纵向坐标 N 的公式中仅包含 l 的偶次幂,且在横向坐标 E 中仅包含奇次幂。系数 $A, \bar{B} \cdots P$,

$Q\cdots$ 为纬度 B 的函数,是长度量纲,且系数中包含了椭球的半径和偏心率参数,即定义测量系统所参考椭球的椭球参数。

纵向和横向坐标 N 和 E 的公式,将椭球正形地转换到平面上,构成该公式的系数 $A, \bar{B} \cdots P$, $Q \cdots$ 是什么函数形式呢?考察参考椭球及其投影平面上的一个简单微分几何示例。如图 7.7 所示,A 点的纬度为 B,经度为 l,假设这两个量微分变化为 $(B + dB)$ 和 $(l + dl)$;dB 意味着沿子午圈长度变化了 MdB,dl 意味着沿平行圈长度变化了 $N\cos Bdl$,变为点 A'。在这些表示中,M 为子午圈曲率半径 $a(1-e^2)(1-e^2\sin^2B)^{-3/2}$,$N$ 为卯酉圈曲率半径 $a(1-e^2\sin^2B)^{-1/2}$,其中 a 为椭球赤道半径,e 为偏心率。

任何投影都可以表示成一般形式,即表示成两个函数:$N = f_1(B, l)$,$E = f_2(B, l)$。若纬度和经度做微小变化 dB 和 dl,相应的纵向和横向坐标 N 和 E 的变化可以表示成两个变量的微分公式

$$dN = \frac{\partial N}{\partial B}dB + \frac{\partial N}{\partial l}dl, \quad dE = \frac{\partial E}{\partial B}dB + \frac{\partial E}{\partial l}dl \tag{7.9}$$

椭球上的微分长度可表示为

$$(dS)^2 = M^2(dB)^2 + N^2\cos^2B(dl)^2 \tag{7.10}$$

相应投影平面上的微分长度 ds 为

$$(ds)^2 = (dN)^2 + (dE)^2 \tag{7.11}$$

从而长度比因子为 ds/dS,代之以 dN 和 dE 的微分公式,可得

$$\left(\frac{ds}{dS}\right)^2 = \left\{\left[\left(\frac{\partial N}{\partial B}\right)^2 + \left(\frac{\partial E}{\partial B}\right)^2\right](dB)^2 + 2\left(\frac{\partial N}{\partial B} \cdot \frac{\partial N}{\partial l} + \frac{\partial E}{\partial B} \cdot \frac{\partial E}{\partial l}\right)(dB)(dl) + \left[\left(\frac{\partial N}{\partial l}\right)^2 + \left(\frac{\partial E}{\partial l}\right)^2\right](dl)^2\right\} / [M^2(dB)^2 + N^2\cos^2B(dl)^2] \tag{7.12}$$

到这里,我们要考虑正形条件,要求上面的公式与 dB 和 dl 的具体数值无关,换句话说,长度比因子与方向无关,这就是正形投影的定义。

现在,来考虑当 $dB = 0$ 和 $dl = 0$ 的两种特殊情况。第一种情况下,则长度比因子的平方简化为

$$\left[\left(\frac{\partial N}{\partial l}\right)^2 + \left(\frac{\partial E}{\partial l}\right)^2\right] / N^2\cos^2B \tag{7.13}$$

第二种情况下,则有

$$\left[\left(\frac{\partial N}{\partial B}\right)^2 + \left(\frac{\partial E}{\partial B}\right)^2\right] / M^2 \tag{7.14}$$

因此,对于正形投影,当然要求这两个分式相等,但只是这一点还不充分。在一般情况下,表达式 $(ds/dS)^2$ 必须与上面两个分式相等,即当 dB 和 dl 都不为零的情况。明显地,除非一般公式中分子中间那一项恒等于零,否则这种情况不成立。所以有了正形投影的第二个条件:

$$\frac{\partial N}{\partial B} \cdot \frac{\partial N}{\partial l} + \frac{\partial E}{\partial B} \cdot \frac{\partial E}{\partial l} = 0 \tag{7.15}$$

第一个条件还可写为:

$$N^2\cos^2B\left[\left(\frac{\partial N}{\partial B}\right)^2 + \left(\frac{\partial E}{\partial B}\right)^2\right] = M^2\left[\left(\frac{\partial N}{\partial l}\right)^2 + \left(\frac{\partial E}{\partial l}\right)^2\right] \tag{7.16}$$

这两个方程就是著名的正形(等角)投影柯西 – 黎曼条件(Canchy-Riemann conditions),确保投

影中任意一点在所有方向上无穷短距离的长度比保持不变。

还要将这些条件应用于纵向和横向坐标 N 和 E 的级数式,然后就可以得到想要的 A, $\bar{B}\cdots P,Q\cdots$ 的公式。如前所述,级数式中的这些因子是纬度 B 的函数,又由于 M 为子午圈曲率半径,根据定义,$MdB = dScosA$。

四个偏微分为

$$\frac{\partial N}{\partial l} = 2Pl + 4Ql^3 + 6Rl^5 \cdots$$

$$\frac{\partial N}{\partial B} = M + \frac{dP}{dB}l^2 + \frac{dQ}{dB}l^4 \cdots$$

$$\frac{\partial E}{\partial l} = A + 3\bar{B}l^2 + 5Cl^4 \cdots$$

$$\frac{\partial E}{\partial B} = \frac{dA}{dB}l + \frac{d\bar{B}}{dB}l^3 + \frac{dC}{dB}l^5 \cdots \quad (7.17)$$

把这些式子代入到柯西–黎曼条件,并把 l 的同次幂项整理到一起,代数运算虽然有一些繁重,但简单明了,条件的前面几项可表示为

$$N^2\cos^2 B \left\{ M^2 + \left[\left(\frac{dA}{dB}\right)^2 + 2M\frac{dP}{dB}\right]l^2 + \left[2\frac{dA}{dB}\cdot\frac{d\bar{B}}{dB} + \left(\frac{dP}{dB}\right)^2 + 2M\frac{dQ}{dB}\right]l^4 \cdots \right\}$$
$$= M^2[A^2 + (6A\bar{B} + 4P^2)l^2 + (10AC + 9\bar{B}^2 + 16PQ)l^4 \cdots] \quad (7.18)$$

且

$$\left(2MP + A\frac{dA}{dB}\right)l + \left(4MQ + 2P\frac{dP}{dB} + 3\bar{B}\frac{dA}{dB} + A\frac{d\bar{B}}{dB}\right)l^3 + \cdots = 0 \quad (7.19)$$

这两个式子是恒等运算,必须要满足 l 的同次幂项以及不包含 l 的项分别相等。由第一个等式,可得 $A = N\cos B$;由第二个等式(l 项),可得 $P = -(A/2M)(dA/dB)$。显然推求 $\bar{B},Q,C\cdots$ 的过程涉及对一系列纬度 B 的函数求微分,例如,$A = a\cos B(1 - e^2\sin^2 B)^{-1/2}$,根据标准求导法则,则 A 的导数为 $-M\sin B$,故可求 $P = +\frac{1}{2}N\sin B\cos B$。然后又可以计算

$$\frac{dP}{dB} = \frac{1}{2}M\cos^2 B\left(\frac{N}{M} - \tan^2 B\right) \quad (7.20)$$

根据第一个等式中的 l^2 项,则可得 $\bar{B} = \frac{1}{6}N\cos^3 B[(N/M) - \tan^2 B]$。在一些公式中出现的比值 N/M,是 $(1 - e^2\sin^2 B)/(1 - e^2)$ 写法的简单表示形式,在很多情况下,可将它取为 1,用于实际计算。

再做一次微分运算,可以由 l^3 项的系数求得 Q;事实上括号里的两内项可以消掉,从而有

$$Q = \frac{1}{24}N\sin B\cos^3 B\left(4\frac{N^2}{M^2} + \frac{N}{M} - \tan^2 B\right) \quad (7.21)$$

在这里,有可能开始考虑对公式做一些简化了。系数 Q 乘以 l^4:N 约为 6.4×10^6 m,且三角函数因式放一起的最大值约为 1.5(在纬度 28°),故 Q 的最大值约为 400 000 m;经度 l 最大约为 1/20,故 Ql^4 项很少会超过 2.5。因此,根据前面的讨论,有 $Q = \frac{1}{24}N(\sin B\cos^3 B)(5 - \tan^2 B)$。

当然,这些简化对于过去使用手工计算和数学用表的时候很有用,有了电子计算机,完全

可以使用完整公式,且在 l^3 项中必须保留 N/M。

最后,为了求得系数 C,有必要计算出 $\mathrm{d}Q/\mathrm{d}B$;经过艰苦细致地完成所有微积分和代数运算,就会得到 C 的完整形式为

$$\frac{1}{120}N\cos^5B\left[4\frac{N^3}{M^3}+\frac{N^2}{M^2}-\left(24\frac{N^3}{M^3}-8\frac{N^2}{M^2}+2\frac{N}{M}\right)\tan^2B+\tan^4B\right] \tag{7.22}$$

或用近似形式

$$\frac{1}{120}N\cos^5B(5-18\tan^2B+\tan^4B) \tag{7.23}$$

实践中,Cl^5 项的最大值为几个厘米。这就足够了!当投影带为 $6°$带时,表示纵向和横向坐标 N 和 E 的级数中,所有其他项的数值很小,无关紧要。

因此,在几度的投影带内,取必要的级数项,计算公式为

$$\begin{aligned}N&=S+\frac{N}{2}\sin B\cos Bl^2+\frac{N}{24}\sin B\cos^3B(5-\tan^2B)l^4\cdots\\E&=N\cos Bl+\frac{N}{6}\cos^3B\left(\frac{N}{M}-\tan^2B\right)l^3+\frac{N}{120}\cos^5B(5-18\tan^2B+\tan^4B)l^5\cdots\end{aligned} \tag{7.24}$$

式中,l 以弧度为单位,$l(弧度)=1''/206\,264.8$。在计算子午线弧长 S 时,必须用 e^2 的幂级数。纬度的一个微小变化 $\mathrm{d}B$ 对应的长度为 $M\mathrm{d}B$,故积分 $\int_0^B M\mathrm{d}B$ 给出了从赤道到纬度 B 的子午线弧长。(参见"子午线弧长")

7.4.2 UTM 投影量的定义及关系

下面给出的椭球面上参量和 UTM 投影平面参量之间关系的定义和解释,仅仅作为一个示例。

在北半球 UTM 投影的图形如图 7.8 所示。

椭球上的直线 $P'_1P'_2$ 一般在投影平面上为曲线,其弯曲总是凹向中央子午线。

线 $P'_1P'_2$ 的方位角表示为 A_{12},它是在 P'_1 点按顺时针量取的 P'_1 点椭球子午线和 P'_2 之间的夹角。

线 P_1P_2 的椭球面长度 S,是在椭球面上点 P_1 和点 P_2 之间的大地线长;曲线 $P'_1P'_2$ 的平面长度 s,是投影点 P'_1 和 P'_2 之间的投影弧长。

线 $P'_1P'_2$ 的平面长度 D,是连接投影点 P'_1 和 P'_2 的弦长。平面弦长 D 和平面曲线长度 s 之间的差别几乎总是可以忽略的。

从点 P'_1 到 P'_2 的平面方位角 T_{12},是在 P'_1 点按顺时针量取的坐标北方向和连接投影点 P'_1 和 P'_2 弦线之间的夹角。

子午线收敛角 γ_1,是在 P'_1 点按顺时针方向量取的真北方向和坐标北方向之间的夹角。

弧到弦的改正 δ_{12},是在线两个端点上弦线和投影弧之间的夹角。点的长度比 m,是平面上该点的微分长度与椭球上相应长度的比值

$$m=\frac{\mathrm{d}s}{\mathrm{d}S} \tag{7.25}$$

由于 UTM 是正形投影,故该比值与微分长度的方位角无关。上面定义的这些参量,这里没有给出其计算公式,但可在各种 UTM 的手册中查到。

7.4.3　长度比和子午线收敛角

有两个投影属性随着点位的不同而不同,在实际中很重要,即长度比和子午线收敛角。长度比为 ds/dS;$(ds/dS)^2$ 的公式,作为纵向和横向坐标 N 和 E 系数的偏微分的函数,在前面"横墨卡托投影中用到的公式"一节中出现过,两个公式任选一个,应用求得的系数 $A,\bar{B}\cdots P,Q\cdots$ 将 B 和 l 项代入,结果为

$$\frac{ds}{dS} = 1 + \frac{1}{2}\frac{N}{M}\cos^2 Bl^2 + \cdots \tag{7.26}$$

由于 $E = N\cos Bl + \cdots$,长度比可以写成更有用的形式:

$$\frac{ds}{dS} = 1 + \frac{1}{2}\frac{E^2}{MN} + \cdots \tag{7.27}$$

若 $E = 200$ km,则长度比可达 1.000 491。多做一些代数运算,可附加 E^4 项,即 $E^4/24N^3M$,但横坐标必须大于 450 km 时,该项对长度比的影响才达百万分之一。

坐标方位角是参照于坐标系纵轴的方向,而真方位角是参照于子午圈的;二者之差就是子午线收敛角,见图 7.8 中的角 LP'_1N(参见"UTM 投影量的定义及关系"一节的图 7.8)。对于过 P_1 点子午圈上的一点,由于经度没有变化,故平面坐标的变化简化为 $dE = (\partial E/\partial B)dB$ 和 $dN = (\partial N/\partial B)dB$。因此,当 $dl = 0$ 时,$\tan\gamma = -(dE/dN)$,也就是,$\tan\gamma = -(\partial E/\partial B)/(\partial N/\partial B)$。应用偏微分的级数式,可得

$$\tan\gamma = l\sin B\left(1 + \frac{1}{3}\frac{N}{M}l^2\cdots\right) \text{ and } \gamma = l\sin B + \frac{1}{3}l^3\sin B\cos^2 B + \cdots \tag{7.28}$$

例如,当 $B = 50°$,$l = 3°$ 时,子午线收敛角为 $2°17'53.28'' + 3.12'' = 2°17'56.4''$。

* * * * * * * * * * *

在我国(中国),高斯 - 克吕格投影在中央子午线上有"真实的"长度比,在中央子午线以外长度比变大。在有些国家,通常将长度比做一个整体的减小,以使投影带边界的长度变形降低约50%。假如,减小因子为 $1/F$,则横坐标 E 的长度比为 $1 + (1/2)(E^2/MN) - 1/F$。例如,若 F 为 2 000,则在约 $E = 202$ km 处,长度比为真实长度比;而在约 $E = 285$ km 处,长度比为 $1/2\,000$,长度比很大。在计算任何长度量之前,可通过减小椭球的半径 a 使减小因子生效,同样,任何测量距离也应当做同样的归算。

8　大地坐标系的建立

8.1　基　准

各个国家,或者在有些情况下是几个国家一起,选择不同的参考椭球。影响这些选择的要素有,椭球的形状和大小,以及椭球的定位。定义了定向、定位以及形状和大小的椭球称为大地基准。在 GPS 定位的背景下,这些不同的大地基准常称之为局部基准,因为与卫星基准相比,局部基准仅适用于地球表面的一个地区或局域范围。

在一般的应用中,有两种主要的卫星基准:精密星历基准和预报星历基准。

8.1.1 局部基准

大地基准或局部基准的定义通常相当主观,其选择仅仅考虑建立的方便。椭球的形状和大小必须首先由选定的长半径 a 和扁率 f 来定义。过去,这些参数的选择往往依赖于历史的发展,有时选择达成协议的国际椭球因为国际椭球体是一致认可的,且经常更新。选定一个控制点,即大地原点,并定义该点相对于椭球的高程,这一定义可以是任意的,也可以根据水准测量的成果导出。水准测量与平均海水面(大地水准面)有关,使用水准测量高程作为大地高意味着定义大地原点上的大地水准面与椭球面重合。大地原点的位置也是用大地纬度和大地经度来定义。一种方法是,简单地采用该点经观测得到的天文纬度和天文经度,下面的椭球面与过该点的水准面平行,换句话说,在大地原点上垂线偏差为零。需要说明一点,要么选择大地方位角,要么选择大地经度,但不能同时选定二者,因为必须要满足拉普拉斯方程。

选择了短轴的定向就完成了基准的定义。一般选择短轴与地球自转轴平行,但是,由于极移,自转轴的方向与时间有关,最方便的定义就是采用国际协议原点极(CIO)。

用这种方法选定的基准一般不是地心的,因为通常没有用到重力观测量来计算大地原点的垂线偏差分量 (ξ, η) 和大地水准面高 (N_0)。利用多普勒观测可以获得有价值的定向信息。

8.1.2 卫星基准

卫星基准以一种与局部基准完全不同的方式定义。局部基准仅仅涉及地球表面的部分区域,与地球的质心,即地心无关,或者只能说是松散相关。前面描述的用于定义局部基准的步骤,跟定义卫星基准所使用的完全不同。卫星参考系是通过给定卫星星历或轨道参数的系统来定义的,这些轨道参数是基于以下要素:包括所采用的许多卫星跟踪站的坐标,所采用的代表地球重力场的地球引力位模型,以及一系列常数。

这些常数包括:

(1)万有引力常数乘以地球质量,GM;

(2)地球相对于瞬时春(秋)分点的自转速率,ω;

(3)光速,c;

(4)跟踪站上用于星历计算的钟差和振荡频率漂移速率。

定义卫星基准的程序参见 Black, H. D., "The Transit System, 1977: performance, plans and potential", Meeting on Satellite Doppler Tracking and its Geodetic Applications at The Royal Society, Oct. 10-11, 1978, London, England,把地球引力位模型包含进来作为卫星基准的主要部分,这意味着系统的坐标原点规定为地球质心。

在定义卫星基准时,参考椭球的概念是不必要的,但是在局部基准定义中它是一个基础。在轨道计算中没有用到椭球定义,但是,通常将椭球与卫星基准联系起来,以便用地理坐标以及直角坐标形式表示位置,这一椭球是用最小二乘法拟合导出的。每个跟踪站由卫星测量导出的半径,减去各点的水准测量高程,得到各点在大地水准面上的对应点,对这些点用最小二乘法拟合出椭球。

8.2 大地基准

大地基准,通常简称为基准,是由中心点(大地原点)相对于所选定的参考椭球中心的位置来定义的。基准通常这样确定,在大地原点测定天文纬度、天文经度和天文方位角,定义参考椭球的形状和大小,并且,最好通过拉普拉斯方程实现椭球自转轴与地球平自转轴平行的条件。若垂线偏差分量和大地水准面差距是绝对的(即,不是相对于参考椭球的),则这一基准是地心基准。由于确定大地基准的大地原点上的这些绝对值很困难,每个国家通常建立自己的相对大地基准。

* * * * * * * * * * *

大地基准通常由五个参数来定义: $a, f; x_0, y_0, z_0$。这里, a 和 f 表示的是参考椭球的长半径和扁率,参考椭球选为旋转椭球; x_0, y_0, z_0 为参考椭球中心相对于地心的直角坐标,地心即地球的质量中心。这个定义预先假定了一个潜在的关于直角 XYZ 的基础坐标系,其 Z 轴与地球的平自转轴重合,且 X 轴通过零子午线,也就是平格林尼治子午线。椭球的自转轴应当与 Z 轴平行。

* * * * * * * * * * *

包含五个参量的参考面:大地原点的纬度和经度,从该点起始的一条线的方位角,以及定义参考椭球所需的两个常数。该参考面构成了水平控制测量计算的基础,在计算中顾及了地球的弯曲。英文中亦作 horizontal control datum; horizontal datum; horizontal geodetic datum。

* * * * * * * * * * *

所需要的地面点的六个参数可写为 $B_0, L_0, h_0, \xi_0, \eta_0, A_0$,它们的作用如下:

(1) B_0, L_0, h_0 为大地原点 P_0 的三个大地坐标,其中 h_0 为正高 H_0 和相对于待定位基准的(相对)大地水准面高 N_0 之和。前面两个参数, B_0, L_0,确定了参考椭球的一条特定法线。由于地方天文系统相对于地球重力场是固定的,一旦给定了 ξ_0 和 η_0,则参考椭球的法线(由 B_0, L_0 确定)相对于地球也随之固定。所选择的 h_0 则固定了参考椭球面低于或高出大地原点($\varphi_0, \lambda_0, H_0$)的高度;在这一阶段,椭球只能绕法线自由转动了。

(2) A_0 为大地方位角,应当满足 $\alpha_0 - A_0 = (\lambda_0 - L)\sin B$,其中, λ_0 和 α_0 相对于 P_0 在大地水准面上的投影。A_0 的选择排除了参考椭球剩下的自由度。

(3) ξ_0, η_0,是 P_0 点上相对于参考椭球的两个(相对)大地水准面垂线偏差分量。

这六个参数等价于六个地心参数,即三个平移参数 X_0, Y_0, Z_0 和三个旋转参数 $\varepsilon_X, \varepsilon_Y, \varepsilon_Z$。地面点六参数相比地心六参数的优点在于,地面点六参数跟地球表面的观测量直接相关联。

* * * * * * * * * * *

认识到地图是画出来的,而位置是相对于某参考基准定义出来的,这一点很重要。在美国使用北美基准(NAD),在日本使用东京基准,在欧洲使用欧洲基准等,当前 GPS 使用 WGS-84。结果,相同的参考点在每一个参考基准下有不同的经纬度坐标,在有些地方,甚至出现了 1/2 km 的明显差别。

图 8.1 中的四幅图帮助想象参考基准的概念以及它们之间的相互联系。我们已经指出,由于密度(重力)差异,地球的形状不规则,图 8.1(a)是不规则地球的一个夸大模型,其表面代表大地水准面,大地水准面定义为平均海水面穿过整个地球表面形成的位置。

为了制作相当准确的地图,需要一个地球表面模型。图 8.1(b)显示了怎样设计这种模型,使得在感兴趣的地区与地球吻合,过去这种地区范围不会超过一个大洲。这一模型包括一个椭球和一个点的位置,称为基准,在该点定义了经度和纬度。这样一个模型很实用,满足在基准邻近地区制作精确地图。

既然现在是应用卫星来测量大地水准面(卫星大地测量学),那么就需要一种不同的基准。如图 8.1(c)表明的那样,一个世界椭球也许不能与地球在某一位置吻合得很好,但它与整个地球是最吻合的,此外,因为涉及许多卫星跟踪站,不存在一个单独的参考基准位置,这些跟踪站的位置定义为确定地球重力场(大地水准面)计算的一部分。WGS-84 椭球就跟 WGS-84 大地水准面吻合得最好。

图 8.1(d)表明必须要用一定的方法把一个基准下的位置与另外一个基准下的坐标联系起来。例如,在东京港获得的卫星定位方位,当画到地方航海图上时,可能会显示轮船恰好在陆地上,原因就是由于图 8.1(d)所示的基准不同。

8.3 坐标变换和基准转换

8.3.1 数学模型

大地基准由参考椭球的参数(长半径 a 和扁率 f)及其相对于地球或大地水准面的位置来确定。这一相对位置通常由大地原点 P_1 上的大地水准面高 N_1 和垂线偏差分量 ξ_1, η_1 给定,也可以用 P_1 点的大地坐标 B_1, L_1, h_1 来代替 ξ_1, η_1, N_1,因为

$$\begin{aligned} \xi_1 &= \varphi_1 - B_1 \\ \eta_1 &= (\lambda_1 - L_1)\cos B_1 \\ N_1 &= h_1 - H_1 \end{aligned} \quad (8.1)$$

一种表面上不同但等价的方法是用参考椭球中心相对于地球中心的直角坐标 x_0, y_0, z_0 来实现。若改变大地基准,即改变参考椭球及其定位,则大地坐标会改变,因此垂线偏差和大地水准面高

$$\begin{aligned} \xi &= \varphi - B \\ \eta &= (\lambda - L)\cos B \\ N &= h - H \end{aligned} \quad (8.2)$$

也会变化。由于有三种不同的方法可固定基准,这些方法可根据以下要素的不同来区分:ξ_1, η_1, N_1,或 B_1, L_1, h_1,或 x_0, y_0, z_0。

从数学上讲,这个问题仅仅是坐标的转换,因为每个大地基准对应一个不同的大地坐标系 B, L, h。

假定参考椭球的中心跟地球的重心不重合,但椭球的短轴与地球的自转轴平行,设想有一个直角坐标系 XYZ,其原点为地球的重心(而非椭球中心),坐标轴的指向同往常一样。令椭球中心相对于这一系统的坐标为 x_0, y_0, z_0,跟前面一样,则大地坐标 (B, L, h) 和直角坐标 (X, Y, Z) 之间的方程很明显必须改化为如下形式

$$X = x_0 + (N + h)\cos B \cos L$$
$$Y = y_0 + (N + h)\cos B \sin L \tag{8.3}$$
$$Z = z_0 + \left(\frac{b^2}{a^2}N + h\right)\sin B$$

这几个方程就是几种重要的坐标变化微分公式的出发点。首先要问的是,若将大地坐标 B,L,h 改变一小量 $\delta B,\delta L,\delta h$,且同时也改变了大地基准,即将参考椭球 (a,f) 及其定位 (x_0,y_0,z_0) 分别改变 $\delta a,\delta f$ 和 $\delta x_0,\delta y_0,\delta z_0$,则直角坐标 X,Y,Z 如何变化。注意 $\delta x_0,\delta y_0,\delta z_0$ 代表参考椭球做小的平移(平行位移),轴向保持跟地球轴向的平行。这一问题的解可通过将式(8.3)微分得到

$$\delta X = \delta x_0 + \frac{\partial X}{\partial a}\delta a + \frac{\partial X}{\partial f}\delta f + \frac{\partial X}{\partial B}\delta B + \frac{\partial X}{\partial L}\delta L + \frac{\partial X}{\partial h}\delta h$$
$$\delta Y = \delta y_0 + \frac{\partial Y}{\partial a}\delta a + \frac{\partial Y}{\partial f}\delta f + \frac{\partial Y}{\partial B}\delta B + \frac{\partial Y}{\partial L}\delta L + \frac{\partial Y}{\partial h}\delta h \tag{8.4}$$
$$\delta Z = \delta z_0 + \frac{\partial Z}{\partial a}\delta a + \frac{\partial Z}{\partial f}\delta f + \frac{\partial Z}{\partial B}\delta B + \frac{\partial Z}{\partial L}\delta L + \frac{\partial Z}{\partial h}\delta h$$

因为,根据泰勒定理,微小的变化可以当作微分处理。

在这些微分方程中,也可以做一些近似。由于扁率 f 很小,可将 $N = a^2/b \sqrt{1 + e'^2\cos^2 B}$ 展开为

$$\begin{aligned}N &= \frac{a^2}{b}(1 + e'^2\cos^2 B)^{-1/2} \\ &= \frac{a^2}{b}\left(1 - \frac{1}{2}e'^2\cos^2 B\cdots\right) \\ &= a(1 + f\cdots)(1 - f\cos^2 B\cdots) \\ &= a(1 + f - f\cos^2 B\cdots)\end{aligned} \tag{8.5}$$
$$N \doteq a(1 + f\sin^2 B) \tag{8.6}$$

从而

$$\frac{b^2}{a^2}N = (1 - 2f\cdots)a(1 + f\sin^2 B\cdots) \doteq a(1 - 2f + f\sin^2 B\cdots) \tag{8.7}$$

因为

$$b = a(1 - f), \quad e'^2 = 2f\cdots \tag{8.8}$$

因此方程(8.3)可近似为

$$X = x_0 + (a + af\sin^2 B + h)\cos B\cos L$$
$$Y = y_0 + (a + af\sin^2 B + h)\cos B\sin L \tag{8.9}$$
$$Z = z_0 + (a - 2af + af\sin^2 B + h)\sin B$$

现在可以组成式(8.4)中的偏导数了,如

$$\frac{\partial X}{\partial a} = (1 + f\sin^2 B)\cos B\cos L \doteq \cos B\cos L \tag{8.10}$$

因为可以忽略系数中的扁率。应用这些系数,且仅用这些系数,意味着,采用了球近似。类似地,作为偏导数,所有系数可以很容易地获得,则方程式(8.4)变为

$$\delta X = \delta x_0 - a\sin B\cos L\delta B - a\cos B\sin L\delta L + \cos B\cos L(\delta h + \delta a + a\sin^2 B\delta f) \quad (8.11a)$$

$$\delta Y = \delta y_0 - a\sin B\sin L\delta B + a\cos B\cos L\delta L + \cos B\sin L(\delta h + \delta a + a\sin^2 B\delta f) \quad (8.11b)$$

$$\delta Z = \delta z_0 + a\cos B\delta B + \sin B(\delta h + \delta a + a\sin^2 B\delta f) - 2a\sin B\delta f \quad (8.11c)$$

这些方程给出了,当椭球的定位(x_0,y_0,z_0)和椭球参数(a,f),以及参考该椭球的大地坐标B,L,h变化时,相应直角坐标X,Y,Z的变化。

8.3.2 大地坐标的变换

可以从方程式(8.11)导出几个重要的坐标转换公式。首先,令P点的空间位置保持不变,即,令

$$\delta X = \delta Y = \delta Z = 0$$

若参考椭球的参数及其定位发生了变化,确定大地坐标B,L,h的变化。

这样,问题就是,设方程式(8.11)左边为零,求解$\delta B,\delta L,\delta h$。为了求$\delta B$,式(8.11a)乘以$-\sin B\cos L$,式(8.11b)乘以$-\sin B\sin L$,式(8.11c)乘以$\cos B$,然后把如此得到的方程加到一起。为求$\delta L$,因子分别为$-\sin L,\cos L$和$0$;为求$\delta h$,因子为$\cos B\cos L,\cos B\sin L$和$\sin B$。结果为

$$\begin{aligned} a\delta B &= \sin B\cos L\delta x_0 + \sin B\sin L\delta y_0 - \cos B\delta z_0 + 2a\sin B\cos B\delta f \\ a\cos B\delta L &= \sin L\delta x_0 - \cos L\delta y_0 \\ \delta h &= -\cos B\cos L\delta x_0 - \cos B\sin L\delta y_0 - \sin B\delta z_0 - \delta a + a\sin^2 B\delta f \end{aligned} \quad (8.12)$$

已经知道,椭球的平移也可以用大地原点上的大地坐标变化$\delta B,\delta L,\delta h$表示,而不是用$\delta x_0,\delta y_0,\delta z_0$表示,这样问题就转化为确定其它点上的大地坐标变化$\delta B,\delta L,\delta h$。

首先,将椭球的平行位移(x_0,y_0,z_0)用给定的$\delta B,\delta L,\delta h$来表示。在式(8.11)中,令(同样因为点在空间中的位置没有改变)$\delta X = \delta Y = \delta Z = 0$,以及$B = B_1, L = L_1, h = h_1$,则有

$$\begin{aligned} \delta x_0 &= a\sin B_1\cos L_1\delta B_1 + a\cos B_1\sin L_1\delta L_1 - \cos B_1\cos L_1(\delta h_1 + \delta a + a\sin^2 B_1\delta f) \\ \delta y_0 &= a\sin B_1\sin L_1\delta B_1 - a\cos B_1\cos L_1\delta L_1 - \cos B_1\sin L_1(\delta h_1 + \delta a + a\sin^2 B_1\delta f) \\ \delta z_0 &= -a\cos B_1\delta B_1 - \sin B_1(\delta h_1 + \delta a + a\sin^2 B_1\delta f) + 2a\sin B_1\delta f \end{aligned} \quad (8.13)$$

将表示平移分量x_0,y_0,z_0的这些表达式代入式(8.12),最后得到

$$\begin{aligned} \delta B &= (\cos B_1\cos B + \sin B_1\sin B\cos\Delta L)\delta B_1 - \sin B\sin\Delta L\cos B_1\delta L_1 + \\ &\quad (\sin B_1\cos B - \cos B_1\sin B\cos\Delta L)\left(\frac{\delta h_1}{a} + \frac{\delta a}{a} + \sin^2 B_1\delta f\right) + \\ &\quad 2\cos B(\sin B - \sin B_1)\delta f \\ \cos B\delta L &= \sin B_1\sin\Delta L\delta B_1 + \cos\Delta L\cos B_1\delta L_1 - \\ &\quad \cos B_1\sin\Delta L\left(\frac{\delta h_1}{a} + \frac{\delta a}{a} + \sin^2 B_1\delta f\right) \\ \frac{\delta h}{a} &= (\cos B_1\sin B - \sin B_1\cos B\cos\Delta L)\delta B_1 + \cos B\sin\Delta L\cos B_1\delta L_1 + \\ &\quad (\sin B_1\sin B + \cos B_1\cos B\cos\Delta L)\left(\frac{\delta h_1}{a} + \frac{\delta a}{a} + \sin^2 B_1\delta f\right) - \\ &\quad \frac{\delta a}{a} + (\sin^2 B - 2\sin B_1\sin B)\delta f \end{aligned} \quad (8.14)$$

其中
$$\Delta L = L - L_1$$

这些公式将任意一点的大地坐标变化 $\delta B, \delta L, \delta h$ 表示为给定点的大地坐标变化 $\delta B_1, \delta L_1, \delta h_1$ 以及参考椭球参数的变化 δa 和 δf，从而将两个不同的大地坐标系联系起来，前提是这两个系统相互之间充分接近，其差别可以认为是线性的。从数学上讲，式(8.14)是微元坐标转换；对于大地测量学家来说，它给出了大地基准变化的影响，跟式(8.12)等价。式(8.12)和式(8.14)都是大地坐标的微元转换；区别仅在于用来确定坐标系，即大地基准的参数不同，在式(8.12)中，坐标系是由 (a,f,x_0,y_0,z_0) 来定义，而式(8.14)中，是由 (a,f,B_1,L_1,h_1) 来定义。

8.3.3 ξ, η, N 的转换

通常，将式(8.14)表示成以垂线偏差分量 ξ 与 η 和大地水准面高 N 为变量的形式。由于自然坐标 φ, λ, H 不受基准转换的影响，保持恒定不变，故由式(8.2)有

$$\begin{aligned} \delta B &= -\delta \xi \\ \delta L \cos B &= -\delta \eta \\ \delta h &= \delta N \end{aligned} \tag{8.15}$$

故式(8.14)变为

$$\begin{aligned} \delta \xi &= (\cos B_1 \cos B + \sin B_1 \sin B \cos \Delta L) \delta \xi_1 - \sin B \sin \Delta L \delta \eta_1 - \\ &\quad (\sin B_1 \cos B - \cos B_1 \sin B \cos \Delta L) \left(\frac{\delta N_1}{a} + \frac{\delta a}{a} + \sin^2 B_1 \delta f \right) - \\ &\quad 2 \cos B (\sin B - \sin B_1) \delta f \\ \delta \eta &= \sin B_1 \sin \Delta L \delta \xi_1 + \cos \Delta L \delta \eta_1 + \\ &\quad \cos B_1 \sin \Delta L \left(\frac{\delta N_1}{a} + \frac{\delta a}{a} + \sin^2 B_1 \delta f \right) \\ \frac{\delta N}{a} &= -(\cos B_1 \sin B - \sin B_1 \cos B \cos \Delta L) \delta \xi_1 - \cos B \sin \Delta L \delta \eta_1 + \\ &\quad (\sin B_1 \sin B + \cos B_1 \cos B \cos \Delta L) \left(\frac{\delta N_1}{a} + \frac{\delta a}{a} + \sin^2 B_1 \delta f \right) - \\ &\quad \frac{\delta a}{a} + (\sin^2 B - 2 \sin B_1 \sin B) \delta f \end{aligned} \tag{8.16}$$

这些表示大地基准转换所产生影响的公式，属于大地测量学中最重要的公式，它们是由包含 Vening Meinesz(1950,1953)在内的几个科学家推导出来的，并以 Vening Meinesz 的名字而著称。早期，也使用由赫尔默特导出的几个形式上相似的公式，但它们是完全基于微分几何原理为基础的，并不适合用于现代大地测量学。

8.3.4 应用

作为例子，把这些公式应用于最重要、最实际的情况，就是局部大地坐标系的绝对定向，或者说是局部大地坐标系与世界大地系的转换。假设已经将三角网或三边网在局部大地基准 $(a', f', \xi'_1, \eta'_1, N'_1)$ 下进行了计算，参照于该系统的量用上标标明，故 ξ'_1, η'_1, N'_1 表示的是大地原点 P_1 的量，可以假定为零或设为其他值。

再假定大地原点上的绝对大地水准面高 N_1 和绝对垂线偏差分量 ξ_1 和 η_1 已知,这几个绝对量值 N,ξ,η 一般是参照于一个不同的椭球(a,f) 的,其中心在地球的重心,量值 a,f,ξ_1,η_1,N_1 则完全确定了这一"世界大地系"。

很容易将局部系统 $(a',f',\xi_1',\eta_1',N_1')$ 转换到世界大地系下。令

$$\begin{aligned} \delta\xi_1 &= \xi_1 - \xi_1' \\ \delta\eta_1 &= \eta_1 - \eta_1' \\ \delta N_1 &= N_1 - N_1' \\ \delta a &= a - a' \\ \delta f &= f - f' \end{aligned} \quad (8.17)$$

对局部系统下的所有点,根据式(8.16)计算变化量 $\delta\xi,\delta\eta,\delta N$,则世界大地系下的相应量 ξ,η,N 为

$$\begin{aligned} \xi &= \xi' + \delta\xi \\ \eta &= \eta' + \delta\eta \\ N &= N' + \delta N \end{aligned} \quad (8.18)$$

世界大地系下的大地坐标则可按下式计算

$$\begin{aligned} B &= B' - \delta\xi \\ L &= L' - \delta\eta\sec B \\ h &= h' + \delta N \end{aligned} \quad (8.19)$$

且地心空间直角坐标 X,Y,Z 可根据 (B,L,h) 和 (X,Y,Z) 的方程关系计算。一个相关的问题是,确定定义局部基准 $(a',f',\xi_1',\eta_1',N_1')$ 的原参考椭球的中心坐标 x_0',y_0',z_0'。由于新基准 (a,f,ξ_1,η_1,N_1),即世界基准,是绝对定位的,故

$$x_0 = y_0 = z_0 = 0$$

可得

$$\begin{aligned} \delta x_0 &= x_0 - x_0' = -x_0' \\ \delta y_0 &= y_0 - y_0' = -y_0' \\ \delta z_0 &= z_0 - z_0' = -z_0' \end{aligned} \quad (8.20)$$

即

$$x_0' = -\delta x_0, \quad y_0' = -\delta y_0, \quad z_0' = -\delta z_0 \quad (8.21)$$

式中,$\delta x_0,\delta y_0,\delta z_0$ 可根据式(8.13)计算。问题得解。

8.4 天文大地坐标系的定向,最佳吻合椭球

通过确定控制网中很多点的垂线偏差和大地水准面高,而不是仅仅确定一个大地原点上的量,可能会改善天文大地坐标系与大地水准面的拟合。

在三维投影方法中,允许椭球相对于大地水准面有三个平移量,且允许椭球参数变化,大地坐标系和全球坐标系之间的平行性应当保持。大地基准参数的相应变化 $\delta\xi_1,\delta\eta_1,\delta N_1,\delta a,\delta f$ 会引起所有控制网点上的椭球坐标、垂线偏差 $\delta\xi,\delta\eta$ 以及大地水准面高 δN 的变化。$\delta\xi,\delta\eta$ 和 δN 可根据 F. A. Vening Meinesz(1950)的垂线偏差投影公式计算(球近似,参见"坐标变换

和基准转换"中式(8.16))。

某特定大地基准下,许多点的垂线偏差 ξ', η' 或大地水准面高 N' 可求得,令

$$\xi = \xi' + \delta\xi, \quad \eta = \eta' + \delta\eta, \quad N = N' + \delta N \tag{8.22}$$

式中,$\delta\xi, \delta\eta$ 和 δN 可按式(8.16)求解(参见"坐标变换和基准变换")。

求最小值的条件

$$\sum (\xi^2 + \eta^2) = \min \tag{8.23}$$

或者,同样有用的条件

$$\sum N^2 = \min \tag{8.24}$$

提供了对大地基准的改正。然后,式(8.16)给出了每一点上 ξ, η, N 的变化。由赫尔默特开发的二维平移方法,用到了归算到椭球面上的量,在这种方法中,允许控制网的平移($\delta\xi_1, \delta\eta_1$),允许控制网有根据拉普拉斯方程得到的相关旋转,允许尺度变化,且允许椭球改变($\delta a, \delta f$)。控制网点上的 $\delta\xi, \delta\eta$ 变化跟基准改正之间的关系可由垂线偏差平移方程式(8.16)计算。应用条件 $\sum (\xi^2 + \eta^2) = \min$。平差垂线偏差带来基准的变化。

利用区域分布的测量(局域方法),J. F. Hayford(1909)和 T. N. Krassowski(1942),还有其他学者,根据垂线偏差的平差确定了椭球。从这一方面,通过采用 Pratt 的均衡补偿理论,Hayford(1909)对美国的观测垂线偏差加入了地形质量影响改正,从而得到了不受不规则质量扰动的椭球参数。克拉索夫斯基应用苏联、美国和欧洲的天文大地观测计算了三轴椭球。参数 $a = 6\ 378\ 245$ m,$f = 1/298.3$ 确定的二轴椭球,是东欧国家进行大地测量的基础。

根据天文大地垂线偏差和大地水准面高计算大地基准,仅仅适用于大洲上,这些方法得到的椭球,在应用方法的相应区域内跟大地水准面的形状和位置吻合得最好(最佳吻合椭球),用这种方式却不能确定平均地球椭球,这些系统的地心位置也不能得到。然而,垂线偏差或大地水准面高在最优吻合的区域内却依然较小,从而,将距离和水平方向归算到椭球时仅有较小的改正数。

由于现在已经有了很多点的地心坐标,还有很好的平均地球椭球,故确定天文大地基准的方法失去了它们的意义,式(8.16)可用于到给定全球基准的转换。

8.5　三维坐标变换

由于多种原因,三维转换更适合用于卫星定位。在概念上,三维转换是典型地全球性的;能同时求解高程和水平位置;数学上是严密的。完全三维转换涉及七个参数,将两个系统下的空间直角坐标联系起来。三个平移参数 $\Delta X, \Delta Y, \Delta Z$ 将两个系统的原点联系起来;三个旋转参数,分别是绕每个坐标轴的旋转量($\omega_X, \omega_Y, \omega_Z$),将两个系统的定向联系起来;以及一个尺度参数($\Delta$),表示两个系统之间的尺度差。

三维转换有两种广泛应用的模型,Bursa-Wolf 模型和 Molodensky-Badekas 模型。

Bursa-Wolf 模型的公式表示如下

$$\begin{pmatrix} X_2 \\ Y_2 \\ Z_2 \end{pmatrix} = \begin{pmatrix} \Delta X \\ \Delta Y \\ \Delta Z \end{pmatrix} + (1 + \Delta) \begin{pmatrix} 1 & -\omega_Z & \omega_Y \\ \omega_Z & 1 & -\omega_X \\ -\omega_Y & \omega_X & 1 \end{pmatrix} \begin{pmatrix} X_1 \\ Y_1 \\ Z_1 \end{pmatrix} \tag{8.25}$$

式中,X_2,Y_2,Z_2为卫星基准下的坐标;X_1,Y_1,Z_1为区域基准下的坐标。

这一模型特别适合于在两个卫星基准之间做比较和转换时使用,不太适合将卫星基准转换到区域基准时使用,因为在计算转换参数时所用到站的地区性(而不是全球分布的),会引起参数间的强相关性。换句话说,在一个有限的区域内,难于区分是由于平移($\Delta X, \Delta Y, \Delta Z$)引起的偏差还是由于旋转($\omega_X, \omega_Y, \omega_Z$)引起的偏差,结果,会出现较大的误差,在区域范围内可以相互抵消,但在区域范围之外,会很快增强为大的基准变换差异。

Molodensky-Badekas 模型则克服了相关性问题,通过将尺度和旋转参数跟某基本点 M 相联系,用这一点的差进行计算。

$$\begin{pmatrix} X_2 \\ Y_2 \\ Z_2 \end{pmatrix} = \begin{pmatrix} \Delta X \\ \Delta Y \\ \Delta Z \end{pmatrix} + \begin{pmatrix} X_M \\ Y_M \\ Z_M \end{pmatrix} + \begin{pmatrix} 1+\Delta & -\omega_Z & \omega_Y \\ \omega_Z & 1+\Delta & -\omega_X \\ -\omega_Y & \omega_X & 1+\Delta \end{pmatrix} \begin{pmatrix} X_1 - X_M \\ Y_1 - Y_M \\ Z_1 - Z_M \end{pmatrix} \qquad (8.26)$$

重新整理,得

$$\begin{pmatrix} X_2 \\ Y_2 \\ Z_2 \end{pmatrix} = \begin{pmatrix} \Delta X \\ \Delta Y \\ \Delta Z \end{pmatrix} + \begin{pmatrix} X_1 \\ Y_1 \\ Z_1 \end{pmatrix} + \begin{pmatrix} \Delta & -\omega_Z & \omega_Y \\ \omega_Z & \Delta & -\omega_X \\ -\omega_Y & \omega_X & \Delta \end{pmatrix} \begin{pmatrix} X_1 - X_M \\ Y_1 - Y_M \\ Z_1 - Z_M \end{pmatrix} \qquad (8.27)$$

美国国防制图局推荐基本点 (X_M, Y_M, Z_M) 取为公共点局域位置的平均

$$\begin{aligned} X_M &= \sum_{i=1}^{n} \frac{X_i}{n} \\ Y_M &= \sum_{i=1}^{n} \frac{Y_i}{n} \\ Z_M &= \sum_{i=1}^{n} \frac{Z_i}{n} \end{aligned} \qquad (8.28)$$

Bursa-Wolf 模型转换参数间的相关性,以及使用 Molodensky-Badekas 模型相关性的降低,可参见 Boucher, C., Investigation on Geodetic Applications of Satellite Doppler Observations for Control Networks, Proceedings of the Second International Geodetic Symposium on Satellite Doppler Positioning, Austin, Texas, January 1979。

Bursa-Wolf 模型和 Molodensky-Badekas 模型中的七个参数可以用最小二乘平差解算。平差用到了点在卫星基准和局域基准下的坐标及其估计方差,通过拟合这两套坐标之间的差值来导出七个转换参数的最小二乘估计,所求得参数的可靠性通常用标准差或方差来表示。

当观察不同参数的量值和标准差时,参数之间尽可能不相关,这一点很重要。当特定参数的标准差等于或大于参数本身时,则有足够的理由可以忽略这些参数。例如,一个 0.07 弧秒的旋转参数,其标准差为 ±0.19 弧秒,则可认为该旋转不显著。若在 Bursa-Wolf 转换中忽略该参数,则由于相关性,其余参数的数值会改变;而在 Molodensky-Badekas 模型中,其余参数将不变。

实际上,经常会发生的一种情况是,两个基准的公共点数不足以精确确定所有七个参数。例如,若有三个公共点,则可以确定所有七个参数,但在解算中仅有两个自由度,在类似这样的情况下,最合理的解就是只确定这三个点的平均平移量 (X,Y,Z),满足于用三参数转换。

这些平移参数也可以用如下的简化模型按最小二乘求解

$$\begin{pmatrix} X_2 \\ Y_2 \\ Z_2 \end{pmatrix} = \begin{pmatrix} X \\ Y \\ Z \end{pmatrix} + \begin{pmatrix} X_1 \\ Y_1 \\ Z_1 \end{pmatrix} \tag{8.29}$$

这一模型对应于将 Bursa-Wolf 或 Molodensky-Badekas 模型中旋转和尺度偏差强制为零。若用于做转换的区域很小的话,则这一模型通常是令人满意的,随着区域范围的增加,未知旋转和尺度变化的影响会变得可观而引起扭曲。

若准备使用三参数转换,希望将模型表示成 $(\Delta B, \Delta L, \Delta h)$,而不是 $(\Delta X, \Delta Y, \Delta Z)$,以便将大地坐标直接从一个系统转换到另一个系统。给定了平移参数 $(\Delta X, \Delta Y, \Delta Z)$、近似坐标 (B, L, h) 和两个椭球的椭球参数,则可用标准或简化 Molodensky 公式计算 $(\Delta B, \Delta L, \Delta h)$ 具体可参见 DMA, "Satellite Records Manual-Doppler Geodetic Point Positioning", DMA TM T-3-52320, Washington, 1976。

由于在 GPS 定位中用到坐标系是空间直角系 (X, Y, Z),因此,通常更方便的方法是,应用 $(\Delta X, \Delta Y, \Delta Z)$ 直接转换一点的空间直角坐标,然后再应用"大地坐标和空间直角坐标"一节给出的公式计算相应的大地坐标 (B, L, h)。在推求转换参数时,所用到的公共点在某一范围内,在使用三参数转换时应该注意,不要使推算超出该范围太远,理想地转换应当仅用于公共点定义范围以内的点(见图 8.2)。

三参数转换实质上假定了两个系统之间不存在旋转和尺度变化。若确实存在旋转和尺度变化,则旋转和尺度影响由三个平移参数吸收,平移参数是通过对公共点上 $\Delta X, \Delta Y, \Delta Z$ 之差取平均算得的。若从这种平均的三参数转换得到的残差合理的话,则可以认为,在公共点的范围之内,充分考虑到了旋转或尺度变化;移到公共点范围之外,这一转换可能不再有效,因为旋转和尺度变化完全会变得很大。为了说明这一点,考虑一种最简单的情况,即仅有一个公共点的情况,从该点求得的 $\Delta X, \Delta Y, \Delta Z$,将会使该点的转换没有误差,即使是两个系统之间存在大的旋转和尺度变化;离开这一点,旋转和尺度变化会变得很显著,故转换不再有效。若有多于一个公共点,则可以检查由平均 $\Delta X, \Delta Y, \Delta Z$ 转换得到的残差,从而确定在公共点所包含的范围之内,转换是否足够准确从而达到了预期目标。

8.6　WGS-84

美国国防部(DoD)建立了 WGS-84 参考系,用来支持其全球性的活动,包括地形图制图、海图制图、定位和导航。更具体地说,DoD 引入 WGS-84 来表示卫星轨道,就是把卫星位置作为时间的函数。因此,WGS-84 广泛用于"绝对"定位,在这里,人们认为卫星轨道足够准确,可以作为定位兴趣点的唯一控制源。尤其是,不依赖使用事先存在的地面点位置坐标作为控制而进行绝对定位,不管怎样,普通用户从来不需要知道跟踪站的位置。

DoD 提供 WGS-84 参考系下的预报轨道和后处理轨道。顾名思义,预报轨道是事先计算的,通过将当前对卫星位置的观测用物理原理外推得到;另一方面,后处理轨道是根据之前对卫星位置的观测计算得到的。后处理轨道比预报轨道更精确,一是因为不涉及对未来的预报,二是因为后处理轨道通常是用很多跟踪站的观测导出的。GPS 预报轨道和卫星钟差参数是由属于 GPS 运行控制部分的位于科罗拉多 Schriever AFB 的美国空军生成。然后美国空军把这些预报值上载到 GPS 卫星,以便 GPS 卫星发射的无线电信号中能包含这些信息,这种预报轨道支持所有涉及 GPS 的实时定位和导航。后处理 GPS 轨道和卫星钟差参数由美国国家影像和制图局(NIMA)生成,NIMA 将这些信息提供到其大地测量和地球物理网站上。许多其他组织也生成后处理 GPS 轨道,通常表示在国际地球参考系(ITRS)的某一具体实现中。

起初的 WGS-84 实现实质上与 NAD83(1986)一致,然而,后来的 WGS-84 实现接近于某 ITRS 实现。由于 GPS 卫星发送 WGS-84 下的预报轨道,故使用广播星历定位点位的用户就自动地获得了与 WGS-84 一致的坐标,因此,使用 GPS 进行实时定位的普及,促进了 WGS-84 更广的应用。尽管应用很广,人们一般并不用 WGS-84 来做高精度定位,因为高精度定位需要用到事先存在的地面点的高精度坐标做控制。例如,许多差分 GPS 技术在计算新点的高精度坐标时,要应用一个或多个预先存在的地面点的已知坐标,来消除某些系统误差。因此,在 WGS-84 支持高精度的定位之前,必须得建立一个相当大规模的控制网,包含精确定位的 WGS-84 地面控制点。

1987 年,DoD 应用海军卫星导航系统(NNSS)或 TRANSIT 的多普勒观测量建立了最初的 WGS-84 参考框架。自从 20 世纪 80 年代中期以来,WGS-84 框架经历了重大演变。1994 年,DoD 引入了一个新的 WGS-84 实现,该实现完全基于 GPS 观测量,而不是多普勒观测。这一新的实现正式叫法为 WGS-84(G730),其中字母 G 代表"GPS",而"730"表示 GPS 周数(从 1994 年 1 月 2 日 0 时开始,UTC 时间),从该周起,NIMA 开始将导出的 GPS 轨道表示在这一框架下。第二个 WGS-84 实现,称为 WGS-84(G873),也是完全基于 GPS 观测实现的,再一次,字母 G 反映了这个事实,"873"指的是从 1996 年 9 月 29 日 0 时(UTC)开始的 GPS 周。最新的 WGS-84 实现,表示为 WGS-84(G1150),是由 NGA(National Geospatial-Intelligence Agency)更新的,GPS 运行控制部分自 2002 年 1 月 20 日起执行该框架。执行改进后的框架,提高了美国空军(USAF)和 NGA GPS 监测站的坐标精度。

WGS-84(G873)的原点、定向和尺度是相对于 15 个 GPS 跟踪站所采用的位置坐标而定的,15 个站中有 5 个是由美国空军维持,10 个由 NIMA 维持。NIMA 选择布站有两个目的,一是在某种程度上增强空军站点的赤道分布,二是使每颗 GPS 卫星尽可能地看到多个跟踪站。

可以预期,在将来,当加入了新的 GPS 跟踪站,或者是现有站的天线迁移或更换时,WGS-84 会做进一步的改进。NGA 致力于采取适当的措施保证最高的质量,并永久保持 WGS-84 的精度。然而,如前所述,大多数地区缺乏可容易参考的控制网,而控制点可以作为控制,采用合适的静态差分 GPS 技术处理载波相位观测量,由控制点可以传递高精度 WGS-84 坐标。关于 WGS-84 的更多信息可通过互联网访问下面的网址获取:http://164.214.2.59/GandG/tr8350_2.html。

8.7 ITRS 及其实现

8.7.1 ITRS 的演化

在 20 世纪 80 年代末,国际地球自转服务(IERS)引入了 ITRS,以支持那些需要高精度位置坐标的科学活动,例如,监测地壳运动和地球自转轴的运动。最初的 ITRS 实现称为国际地球参考框架 1988(ITRF88)。相应的,IERS 发布了几百个站构成的全球网的位置和速率,IERS 在几个合作机构的帮助下,应用几种高精度大地测量技术求得了这些位置和速率,这些技术包括 GPS、VLBI、SLR、LLR 和 DORIS（卫星集成多普勒定轨和无线电定位）。自从引入了 ITRF88，大约每年 IERS 都会开发新的 ITRS 实现（ITRF89、ITRF90、……、ITRF2005）借以发布以前已有站的改进位置和速率，以及上一个实现开发后所建立的新站的坐标和速率。每一次新的实现不仅包括了至少又一年的数据，而且还包括对地球动态特性的最新成果。ITRF96 框架由 508 个站的坐标和速率定义，这 508 个站散布在全球 290 个站点上。要知道，一个特定的站点可能包括一个或多个并址设备，采用多种空间技术（例如，GPS、VLBI、SLR、LLR 以及 DORIS）。ITRS 的精度和严谨，使它在那些从事定位活动人群中的受欢迎程度正稳步增长。

此外，ITRS 通过发布控制点的速率和坐标，是第一个直接处理板块构造和其他形式地壳运动的主要国际参考系。为了理解对速率的需求，来考虑板块构造理论。根据这一理论，地球的外壳包括大约 20 个本质上是刚性的板块，且这些板块相互之间主要是横向运动，就像贮水池中几个巨大的冰片。不同板块上点与点之间的相对运动，在某些情况下有 150 mm/a 那么大，这一量值用 GPS 和其他现代定位技术很容易探测到。

考虑到每一个构造板块都在做相对于其他板块的运动，有人会问怎样才能将地壳速率用绝对量表示。负责 ITRS 的人员现在这样来处理这一难题，即假定地球表面作为一个整体，平均起来相对于地球内部没有移动，换句话说，ITRS 开发者认为地球外壳的总角动量为零，因此，与任何一个板块运动相关的角动量，都由跟剩余板块运动相关的联合角动量来平衡。所以，按照 ITRS 对绝对运动的定义，北美板块上的点通常以可测量的速率做水平运动。特别的，在美国 48 个毗连的州上，ITRF96 的水平速率量值在 10~20 mm/a 之间，此外，在阿拉斯加和夏威夷州，ITRF96 的水平速率量值要更大一些。

相比之下，NAD83 参考系对板块运动的处理基于以下假设，即北美板块作为一个整体，平均起来相对于地球内部没有运动。因此，北美板块上的点相对于 NAD83 一般没有水平运动速率，除非点位位于接近板块边界的地方（加利福尼亚州、俄勒冈州、华盛顿州，以及阿拉斯加），或者是受到其他变形过程（火山/岩浆活动，冰期回弹等）的影响。然而，NAD83 参考系，并没有特殊照顾到完全位于另外一个板块上的某些美国的地区，例如，在夏威夷，假定太平洋板块不运动来定义的 NAD83 点位坐标，这种方法对于仅仅在夏威夷地区从事定位活动的人员来说是方便的，但是，对于涉及确定夏威夷点位相对于北美洲点位定位的人员来说，这种方法则引入了一层复杂性。

在地壳运动领域，给定了位置坐标没有指定这些坐标的历元是不恰当的，历元也就是这些坐标对应的日期。因此，ITRF96 位置通常指定历元日期为 1997 年 1 月 1 日（常以年为单位表示为 1997.0）。为了计算另一个时刻 t 的坐标，需要应用公式

$$x(t) = x(1997.0) + v_x \cdot (t - 1997.0) \tag{8.30}$$

求 $y(t)$ 和 $z(t)$ 的公式与此类似，这里，$x(t)$ 表示点位在时刻 t 的 x 坐标，$x(1997.0)$ 表示点位在 1997 年 1 月 1 日的 x 坐标，v_x 表示点位速率的 x 分量。

8.7.2 ITRF 的背景

空间大地测量学的终极目标之一就是尽可能准确地估计地面点的点位位置，与此同时点位位置既不能观测，也不是绝对量，只能相对于一定的参考来确定。"地球参考系统（TRS）"就是满足理想定义的数学参考系，点位位置可以在其中表达。然而，要确定点位位置还需要与数学参考系相联系的观测手段，称之为"地球参考框架（TRF）"，即 TRS 的物理实现，它是应用

空间大地测量技术的观测成果实现的。

"系统"和"框架"之间的区别是很微妙的。前者相对来说是不变的,不能直接使用的;而后者是可以直接使用的,并且是可以不断改善的。在20世纪80年代,天文学和大地测量学界对参考系和参考框架的基本概念有过广泛的讨论。自20世纪80年代起,空间大地测量技术的应用有效地改善了地球表面的定位精度:从最初的分米级发展到现在的厘米级甚至是毫米级。

然而,每一种技术以及数据分析都定义和实现了各自的TRS,因而存在大量的TRF,如果相互进行比较,会发现他们之间存在系统差异。为此,国际大地测量和地球物理联合会(IUGG)以及国际大地测量协会(IAG)决定所有的地球科学应用采取唯一的TRS,称之为国际地球参考系(ITRS)(Geodesist's Handbook,1992)。关于ITRS的一般描述,参见McCarthy(1996)。ITRS的原点定义为包括海洋和大气的全球质量中心。按照IAU/IUGG的决议(1991)(参见McCarthy(1992)的附录),ITRS的长度单位采用米(SI),这一尺度与地心坐标时(TCG)一致。采用TCG保证地球坐标系下所有物理量的量值与质心系(或其他行星中心坐标系)下的取值相同。地心坐标时TCG的平均变率与观测者位于地心时(移去地球)的原时平均变率相同,而地球时(TT)的平均变率与观测者位于大地水准面上时的原时平均变率相同,这两种时间尺度的变率之差为($TCG - TT \approx 0.7 \times 10^{-9}$)。ITRS定向与国际时间局(BIH)在1984.0的定向一致;定向时间演化是通过相对于全球的水平构造运动无绝对旋转条件来保证。

实现ITRF的基本思想是:各个分析中心应用空间大地测量技术的观测成果,包括VLBI、LLR及SLR、GPS、DORIS的成果,计算测站位置(和速率),把这些测站位置(和速率)联合平差得到ITRF。已经证实,综合了各种技术长处的联合平差方法对生成全球参考框架是有效的。

ITRF的历史可以追溯到1984年,那时应用由VLBI、LLR、SLR和Doppler/TRANSIT观测分析得到的站坐标,第一次建立了一个地球参考框架TRF(称为BTS84)。BTS84是在BIH的活动框架下实现的,BIH是当时一项国际计划MERIT(Monitoring of Earth Rotation and Intercomparison of Techniques)的协调中心。后来又实现了三个后续的BTS(分别是BTS85、BTS86、BTS87),BTS系列以BTS87结束,这是因为在1988年IUGG和IAU联合成立了IERS。

IERS的第一个ITRS实现是ITRF88,此后,共实现并发布了10个版本的ITRF,每个版本的ITRF都取代先前的版本。为了得到最优的联合平差解以生成ITRF,对数据处理技术进行了不断地重大改进。

考虑到ITRF在大地测量和地球物理领域的应用非常广泛,ITRF2000在质量、站网和基准定义等方面做了改进。ITRF2000反映了空间大地测量技术的真实质量,不受任何外部约束的影响;包含了VLBI、LLR、SLR、GPS和DORIS技术(以前ITRF版本通常用到的技术)观测的核心站,以及区域GPS网作为加密;为了保证定向时间演化的稳定性,ITRF2000的定向速率通过选择一组高质量的站点实施定义。

以往的ITRF版本是联合处理全球长期解得到的,与以往版本不同,ITRF2005应用站坐标和每天地球定向参数(EOPs)的时间序列(卫星技术是每周一套解,VLBI技术是24小时一个时段的解)作为输入数据。使用站坐标时间序列的优点在于,允许监测测站的非线性运动和不连续性,检查框架物理参数的时变行为,框架物理参数即原点和尺度。

8.7.3 ITRF 基准定义的背景

从大地测量的观点看,通过 TRF 实现 TRS 需要 14 个参数:三个平移(原点)参数,一个尺度参数,三个旋转(定向),以及这七个参数相应的速率参数。考虑到这 14 个参数可以看作是两个 TRF 之间的相对值,其选择应当满足所采用的 TRS 定义。然而,空间大地测量观测不包含完全实现 TRS 的所有必要信息。卫星技术对地球质量中心(天然的 TRF 原点)敏感,而 VLBI 却不敏感(其 TRF 的原点通过约束条件主观定义的);尺度定义依赖于对某些物理参数的建模;TRF 的定向(不能由任何技术观测)是通过特定的约束主观给定,或者按照惯例定义。

动力学技术(LLR、SLR、GPS、DORIS)轨道的参考点是地球质量中心,然而,由于受到地球内部和表层质量重新分配的多种地球动力学过程影响,引起地球质心位置相对于地球表面随时间而变化。换句话说,由这种效应引起跟踪站的地心的运动,称为"地心运动",可能包含周期性分量和长期性分量。目前,大多数分析中心都没有在模型中包含这种效应,因此,这样定义的 TRF 原点实际上与所采用观测量这段时间内质心的平均位置一致。事实上,目前动力学技术精确测量地心运动的能力有限。

当生成一种联合的 TRF(如 ITRF)时,可以采用不同的选择来具体确定 14 个基准参数。平移、尺度、定向等 7 个参数应当选为某一历元的数值,而尺度速率和平移速率的选择则依赖于参与联合平差的各个 TRF 相对速率的显著性。

卫星 TRF 之间的平移速率严重依赖于网的结构、轨道及所用的观测,又受地心长期运动的影响。尺度速率受测站垂直运动和其他模型的影响,如对流层模型,还受与具体技术有关效应的影响,如 VLBI、GPS、DORIS 的天线相关效应,SLR 与测站有关的测距偏差等。尽管定向速率可以主观选择,但由于板块构造运动,定向速率的选择应具有地球物理含义。

因此,TRS 的实现应当考虑地壳运动和地壳形变,因为空间大地测量观测站正是位于这样的形变地壳上,与此同时,由于空间大地测量技术利用了天体的位置和运动,如卫星和 VLBI 观测的河外射电源,因而对于每个单独(每种技术)的 TRS 实现,自然应当确保天球参考框架(CRF)、TRF 和 EOP(连接 CRF 和 TRF)之间的一致性。因此,ITRF,作为 IERS 的三个全球参考之一,应当与 ICRF 和 EOP 序列是一致的。

此外,地球自转理论的基本方程都和这样一个坐标系相关联,即在这一坐标系中,地球的任何运动和变形都不影响地球绝对角动量。为实现这一目标,一种普遍接受的方法是应用 Tisserand 系统,即由动能最小定义的平均轴:

$$T = \frac{1}{2} \int_C V^2 \mathrm{d}m \qquad (8.31)$$

式中,$\mathrm{d}m$ 为质元;V 是其速度。积分区域 C 最直观的选择是整个地壳,如上所述,因为大地测量观测站是建立在地壳上,地壳运动和形变信息都包含在观测之中。

定义地壳最小动能 T 的目的之一是使地壳的全球运动最小化,因为地壳全球运动会影响到 EOP。当然,更深层的地球动力学过程,如地幔对流和核幔边界等,也会影响地球自转。最小化 T 使得线动量 p 和角动量 h 为零

$$p = \int_C V \mathrm{d}m = 0 \qquad (8.32)$$

$$h = \int_C X \times V \mathrm{d}m = 0 \qquad (8.33)$$

式中,X 为 $\mathrm{d}m$ 的位置矢量。

式(8.32)和式(8.33)分别从理论上定义了 TRF 的平移速率和定向旋转速率。然而尺度速率在这种方法中看不到,它跟由空间大地测量技术估计的物理量有关。

考虑到卫星(动力学)技术能估计地心运动,因而站位置参照于地球质心,地球质心自然作为 TRF 的原点,故不再需要式(8.32),同时,式(8.33)则表示对地壳无整体旋转条件(NNR)的定义。

严格应用式(8.33)需要已知地壳密度以及地壳厚度,而这两者是随空间而变化的,特别是在海洋和大洲交界处,因而计算式(8.33)的积分值时会有困难。相反,而是使用一种求和方法,在一定数目的测站上求和,或者在代表岩石圈的构造板块上求和。这两种方法(基于测站和基于板块)的每一种都会得到 NNR 条件的不同实现。

基于观测测站实现 NNR 条件的方法需要仔细的审查。参与定义的测站应当满足的最低标准,比如,要有足够长的观测时间(比方说至少三年),要在整个地球表面最优分布。此外,在这种方法中,使用式(8.33)需要客观选择(或定界)跟每个测站相关联的每个面元,即应当考虑每个测站的地球物理背景。

从 ITRF94 开始,将单个解集的方差-协方差矩阵包含到全球联合平差。那时,ITRF94 的基准定义如下:原点(三个平移分量),由某些 SLR 解和 GPS 解的加权平均定义;尺度,由 VLBI、SLR 和 GPS 解的加权平均定义,为了满足 IUGG 和 IAU 将尺度定义在 TCG(地心坐标时)时间框架下而不是 TT(地球时)时间框架下的要求,加入 0.7×10^{-9} 的改正,因为各个分析中心采用 TT;定向,与 ITRF92 对准;定向时间演化,通过 7 个转换参数速率将 ITRF94 速率场与 NNR-NUVEL-1A 模型对准来定义。

ITRF96 与 ITRF94 对准,ITRF97 与 ITRF96 对准,用 14 个转换参数。对提交 ITRF2000 的各个解集的分析表明,相对于大多数 SLR 解,ITRF97 表现在 Z 轴的平移(和速率),同样尺度也有类似现象。

有必要定义 ITRF2000 的基准,使得它与最新的空间大地测量数据一致,决定采用如下的基准定义:

(1)尺度因子及其速率,由 VLBI 解和最一致的 SLR 解的加权平均定义。与 ITRF97 尺度表示在 TCG 框架下不同,按照 ITRF2000 工作组的建议(2000 年 11 月),ITRF2000 的尺度表示在 TT 框架下。采用这一决议是为了满足各空间大地测量分析中心为 ITRF2000 提交数据的需要,因为各个分析中心使用的时间尺度与 TT 框架一致。

(2)平移参数及其速率,由最一致的 SLR 解加权平均定义。

(3)定向,在 1997.0 与 ITRF97 的定向一致;定向速率,使得相对于 NNR-NUVEL-1A 无绝对旋转速率。注意,定向及其速率的定义是基于所选择的一组高质量的大地测量站定义的,要满足如下原则:连续观测至少 3 年;坐落在构造板块的稳定地区且远离变形区域;速度误差(ITRF2000 联合平差的结果)小于 3 mm/a;在至少 3 个不同解集中,速度残差小于 3 mm/a。

* * * * * * * * * * *

ITRF2005 的原点、尺度、定向及其时变指定如下:

原点:ITRF2005 的原点这样来定义,历元 2000.0,使得 ITRF2005 和 ILRS 的 SLR 时间序列之间的平移参数和平移速率为零。

尺度:ITRF2005 的原点这样来定义,历元 2000.0,使得 ITRF2005 和 IVS 的 VLBI 时间序列之间的尺度因子和尺度速率为零。选择 VLBI 尺度定义 ITRF2005 的尺度是因为:在提交的时间序列中,全部的 VLBI 观测历史的可用性最长(26 年,而 SLR 为 13 年);在 ILRS 的尺度中观测到了非线性现象(不连续性)。

定向:ITRF2005 的定向这样来定义,历元 2000.0,使得 ITRF2005 和 ITRF2000 之间的旋转参数和旋转速率参数为零,这两个条件是在一组 70 个核心站上实施的。

ITRF2005 的原点这样来定义,使 ITRF2005 相对于地球质心的平移参数和平移速率为零,地球质心取为跨度为 13 年观测的 SLR 时间序列的平均。其尺度定义为,相对于跨度为 26 年观测的 VLBI 时间序列,ITRF2005 的尺度和尺度速率为零。应用 70 个高质量大地测量站,使 ITRF2005 的定向(在历元 2000.0)和定向速率与 ITRF2000 匹配。ITRF2005 原点(在历元 2000.0)及其速率相对于 ITRF2000 一致性水平,估计沿 X、Y 和 Z 轴方向分别为 0.1,0.8,5.8 mm 和 0.2,0.1,1.8 mm/a。

* * * * * * * * * * * *

ITRF2008 采用与 ITRF2005 相同的策略,以四种空间大地测量技术的重新平差解为基础,是一种更精化的解,这四种技术是 VLBI、SLR、GPS 和 DORIS,观测跨度分别为 29 年、26 年、12.5 年和 16 年。

ITRF2008 由 580 个台站上的 934 个测站组成,在南(117 个台站)北(463 个台站)半球分布不均衡。共有 105 个并址站,其中 91 个台站的联结测量用于 ITRF2008 的联合平差。遗憾的是,当前并不是所有并址站上的观测设备都在运行,例如,在具有 4 种技术的 6 个台站上,目前只有 2 个台站是完全运转的:南非的 Hartebeesthoek,以及美国马里兰州的 Greenbelt。

ITRF2008 由以下框架参数规定:

(1) 原点:ITRF2008 的原点这样来定义,历元 2005.0,相对于 ILRS 的 SLR 时间序列平移参数和平移速率为零。

(2) 尺度:ITRF2008 的尺度这样来定义,历元 2005.0,相对于 VLBI 和 SLR 时间序列的平均尺度和尺度速率,尺度因子和尺度速率为零。

(3) 定向:ITRF2008 的定向这样来定义,历元 2005.0,ITRF2008 和 ITRF2005 之间的旋转参数和旋转速率参数为零,这两个条件是在一组位于 131 个台站上的 179 个参考站上实施的。